P9-CKD-647

Cracking the

AP®

PHYSICS 2 EXAM

2017 Edition

by the Staff of The Princeton Review

PrincetonReview.com

Penguin
Random
House

The Princeton Review
24 Prime Parkway, Suite 201
Natick, MA 01760
E-mail: editorialsupport@review.com

Published in the United States by Penguin Random House LLC,
New York, and in Canada by Random House of Canada, a division
of Penguin Random House Ltd., Toronto.

ISBN: 978-1-101-91996-5
eBook ISBN: 978-1-101-92022-0
ISSN: 2380-3290

Editor: Meave Shelton
Production Editors: Lee Elder and Liz Rutzel
Production Artist: Craig Patches

Printed in the United States of America on
partially recycled paper.

10 9 8 7 6 5 4 3 2 1

2017 Edition

Editorial
Robert Franek, Senior VP, Publisher
Casey Cornelius, VP Content Development
Mary Beth Garrick, Director of Production
Selena Coppock, Managing Editor
Meave Shelton, Senior Editor
Colleen Day, Editor
Sarah Litt, Editor
Aaron Riccio, Editor
Orion McBean, Editorial Assistant

Random House Publishing Team
Tom Russell, Publisher
Alison Stoltzfus, Publishing Manager
Jake Eldred, Associate Managing Editor
Ellen Reed, Production Manager

ACKNOWLEDGMENTS

Special thanks to Peter Vaughan for his contributions to the 2017 edition of this book.

Contents

Register Your

1 Go to **PrincetonReview.com/cracking**

2 You'll see a welcome page where you can register your book using the following ISBN: 9781101919965.

3 After placing this free order, you'll either be asked to log in or to answer a few simple questions in order to set up a new Princeton Review account.

4 Finally, click on the "Student Tools" tab located at the top of the screen. It may take an hour or two for your registration to go through, but after that, you're good to go.

If you are experiencing book problems (potential content errors), please contact EditorialSupport@review.com with the full title of the book, its ISBN number (located above), and the page number of the error. Experiencing technical issues? Please e-mail TPRStudentTech@review.com with the following information:

- your full name
- e-mail address used to register the book
- full book title and ISBN
- your computer OS (Mac or PC) and Internet browser (Firefox, Safari, Chrome, etc.)
- description of technical issue

Book Online!

Once you've registered, you can...

- Take a full-length practice PSAT, SAT, and ACT

- Get valuable advice about the college application process, including tips for writing a great essay and where to apply for financial aid

- Sort colleges by whatever you're looking for (such as Best Theater or Dorm), learn more about your top choices, and see how they all rank according to *The Best 380 Colleges*

- Access comprehensive study guides and a variety of printable resources, including bubble sheets for the practice tests in the book as well as important equations and formulas

- Check to see if there have been any updates made to this edition or any late-breaking news releases from the College Board.

Look For These Icons Throughout The Book

 Online Articles

 Study Break

 Proven Techniques

 More Great Books

The
Princeton
Review®

Part I
Using This Book to Improve Your AP Score

- Preview: Your Knowledge, Your Expectations
- Your Guide to Using This Book
- How to Begin

PREVIEW: YOUR KNOWLEDGE, YOUR EXPECTATIONS

Your route to a high score on the AP Physics 2 Exam will depend on how you plan to use this book. Please respond to the following questions.

1. Rate your level of confidence about your knowledge of the content tested by the AP Physics 2 Exam.

    A. Very confident—I know it all
    B. I'm pretty confident, but there are topics for which I could use help
    C. Not confident—I need quite a bit of support
    D. I'm not sure

2. If you have a goal score in mind, circle your goal score for the AP Physics 2 Exam.

    5   4   3   2   1   I'm not sure yet

3. What do you expect to learn from this book? Circle all that apply to you.

    A. A general overview of the test and what to expect
    B. Strategies for how to approach the test
    C. The content tested by this exam
    D. I'm not sure yet

YOUR GUIDE TO USING THIS BOOK

This book is organized to provide as much—or as little—support as you need, so you can use this book in whatever way will be most helpful for improving your score on the AP Physics 2 Exam.

*   The remainder of **Part I** will provide guidance on how to use this book and help you determine your strengths and weaknesses.

*   **Part II** of this book contains your first practice test, answers and explanations, and a scoring guide. (Bubble sheets can be found in the very back of the book for easy tear-out.) This is where you should begin your test preparation in order to realistically determine:
    o   your starting point right now
    o   which question types you're ready for and which you might need to practice
    o   which content topics you are familiar with and which you will want to carefully review
    Once you have nailed down your strengths and weaknesses with regard to this exam, you can focus your preparation and be efficient with your time.

- **Part III** of this book will:
  - provide information about the structure, scoring, and content of the AP Physics 2 Exam
  - help you to make a study plan
  - point you toward additional resources

- **Part IV** of this book will explore the following strategies:
  - how to attack multiple-choice questions
  - how to approach free-response questions
  - how to manage your time to maximize the number of points available to you

- **Part V** of this book covers the content you need for the AP Physics 2 Exam.

- **Part VI** of this book contains Practice Test 2, its answers and explanations, and a scoring guide. (Bubble sheets can be found in the very back of the book for easy tear-out.) If you skipped Practice Test 1, we recommend that you do both (with at least a day or two between them) so that you can compare your progress between the two. Additionally, this will help to identify any external issues: if you get a certain type of question wrong both times, you probably need to review it. If you only got it wrong once, you may have run out of time or been distracted by something. In either case, this will allow you to focus on the factors that caused the discrepancy in scores and to be as prepared as possible on the day of the test.

You may choose to use some parts of this book over others, or you may work through the entire book. This will depend on your needs and how much time you have. Let's now look at how to make this determination.

Equation Sheet Sidebars

Throughout the content review chapters of this book you'll see the words "Equation Sheet" pop up in the sidebars. These words will be seen next to physics equations that are featured on the equation sheet that the College Board gives you during your AP Physics 2 Exam. So these are equations that are extremely important for you to be familiar with.

HOW TO BEGIN

1. **Take Practice Test 1**

    Before you can decide how to use this book, you need to take a practice test. Doing so will give you insight into your strengths and weaknesses, and the test will also help you make an effective study plan. If you're feeling test-phobic, remind yourself that a practice test is a tool for diagnosing yourself—it's not how well you do that matters but how you use information gleaned from your performance to guide your preparation.

    So, before you read further, take AP Physics 2 Practice Test 1 starting at page 9 of this book. Be sure to do so in one sitting, following the instructions that appear before the test.

2. **Check Your Answers**

    Using the answer key on page 28, count how many multiple-choice questions you got right and how many you missed. Don't worry about the explanations for now, and don't worry about why you missed questions. We'll get to that soon.

3. **Reflect on the Test**

    After you take your first test, respond to the following questions:

    - How much time did you spend on the multiple-choice questions?

    - How much time did you spend on each free-response question?

    - How many multiple-choice questions did you miss?

    - Do you feel you had the knowledge to address the subject matter of the essays?

    - Do you feel you wrote well-organized, thoughtful essays?

4. **Read Part III of this Book and Complete the Self-Evaluation**

    Part III will provide information on how the test is structured and scored. It will also set out areas of content that are tested.

    As you read Part III, re-evaluate your answers to the questions above. At the end of Part III, you will revisit the questions on the previous page and refine your answers to them. You will then be able to make a study plan, based on your needs and time available, that will allow you to use this book most effectively.

5. **Engage with Parts IV and V as Needed**

Notice the word *engage*. You'll get more out of this book if you use it intentionally than if you read it passively, hoping for an improved score through osmosis.

Strategy chapters will help you think about your approach to the question types on this exam. Part IV will open with a reminder to think about how you approach questions now and then close with a reflection section asking you to think about how/whether you will change your approach in the future.

Content chapters are designed to provide a review of the content tested on the AP Physics 2 Exam, including the level of detail you need to know and how the content is tested. You will have the opportunity to assess your mastery of the content of each chapter through test-appropriate questions and a reflection section.

6. **Take Practice Test 2 and Assess Your Performance**

Once you feel you have developed the strategies you need and gained the knowledge you lacked, you should take Practice Test 2. You should do so in one sitting, following the instructions at the beginning of the test.

When you are done, check your answers to the multiple-choice sections. See if a teacher will read your essays and provide feedback.

Once you have taken the test, reflect on what areas you still need to work on, and revisit the chapters in this book that address those deficiencies. Through this type of reflection and engagement, you will continue to improve.

7. **Keep Working**

After you have revisited certain chapters in this book, continue the process of testing, reflecting, and engaging with the next practice test in this book. Consider what additional work you need to do and how you will change your strategic approach to different parts of the test. You can continue to explore areas that can stand to be improved and engage in those areas right up to the day of the test. As we will discuss in Part III, there are other resources available to you, including a wealth of information online at AP Students.

Part II
Practice Test 1

- Practice Test 1
- Practice Test 1: Answers and Explanations

Practice Test 1

AP® Physics 2 Exam

SECTION I: Multiple-Choice Questions

DO NOT OPEN THIS BOOKLET UNTIL YOU ARE TOLD TO DO SO.

At a Glance

Total Time
90 minutes
Number of Questions
50
Percent of Total Grade
50%
Writing Instrument
Pen required

Instructions

Section I of this examination contains 50 multiple-choice questions. Fill in only the ovals for numbers 1 through 50 on your answer sheet.

CALCULATORS MAY BE USED IN BOTH SECTIONS OF THE EXAMINATION.

Indicate all of your answers to the multiple-choice questions on the answer sheet. No credit will be given for anything written in this exam booklet, but you may use the booklet for notes or scratch work. After you have decided which of the suggested answers is best, completely fill in the corresponding oval on the answer sheet. Give only one answer to each question. If you change an answer, be sure that the previous mark is erased completely. Here is a sample question and answer.

Sample Question   Sample Answer

Chicago is a   Ⓐ ● Ⓒ Ⓓ
(A) state
(B) city
(C) country
(D) continent

Use your time effectively, working as quickly as you can without losing accuracy. Do not spend too much time on any one question. Go on to other questions and come back to the ones you have not answered if you have time. It is not expected that everyone will know the answers to all the multiple-choice questions.

About Guessing

Many candidates wonder whether or not to guess the answers to questions about which they are not certain. Multiple choice scores are based on the number of questions answered correctly. Points are not deducted for incorrect answers, and no points are awarded for unanswered questions. Because points are not deducted for incorrect answers, you are encouraged to answer all multiple-choice questions. On any questions you do not know the answer to, you should eliminate as many choices as you can, and then select the best answer among the remaining choices.

GO ON TO THE NEXT PAGE.

ADVANCED PLACEMENT PHYSICS 2 TABLE OF INFORMATION

CONSTANTS AND CONVERSION FACTORS	
Proton mass, $m_p = 1.67 \times 10^{-27}$ kg	Electron charge magnitude, $e = 1.60 \times 10^{-19}$ C
Neutron mass, $m_n = 1.67 \times 10^{-27}$ kg	1 electron volt, 1 eV $= 1.60 \times 10^{-19}$ J
Electron mass, $m_e = 9.11 \times 10^{-31}$ kg	Speed of light, $c = 3.00 \times 10^8$ m/s
Avogadro's number, $N_0 = 6.02 \times 10^{23}$ mol^{-1}	Universal gravitational constant, $G = 6.67 \times 10^{-11}$ m^3/kg·s^2
Universal gas constant, $R = 8.31$ J/(mol·K)	Acceleration due to gravity at Earth's surface, $g = 9.8$ m/s^2
Boltzmann's constant, $k_B = 1.38 \times 10^{-23}$ J/K	

1 unified atomic mass unit,	1 u $= 1.66 \times 10^{-27}$ kg $= 931$ MeV$/c^2$
Planck's constant,	$h = 6.63 \times 10^{-34}$ J·s $= 4.14 \times 10^{-15}$ eV·s
	$hc = 1.99 \times 10^{-25}$ J·m $= 1.24 \times 10^3$ eV·nm
Vacuum permittivity,	$\varepsilon_0 = 8.85 \times 10^{-12}$ C^2/N·m^2
Coulomb's law constant,	$k = 1/4\pi\varepsilon_0 = 9.0 \times 10^9$ N·m^2/C^2
Vacuum permeability,	$\mu_0 = 4\pi \times 10^{-7}$ (T·m)/A
Magnetic constant,	$k' = \mu_0/4\pi = 1 \times 10^{-7}$ (T·m)/A
1 atmosphere pressure,	1 atm $= 1.0 \times 10^5$ N/m^2 $= 1.0 \times 10^5$ Pa

UNIT SYMBOLS							
	meter,	m	mole,	mol	watt,	W	farad,	F
	kilogram,	kg	hertz,	Hz	coulomb,	C	tesla,	T
	second,	s	newton,	N	volt,	V	degree Celsius,	°C
	ampere,	A	pascal,	Pa	ohm,	Ω	electron volt,	eV
	kelvin,	K	joule,	J	henry,	H		

PREFIXES		
Factor	Prefix	Symbol
10^{12}	tera	T
10^9	giga	G
10^6	mega	M
10^3	kilo	k
10^{-2}	centi	c
10^{-3}	milli	m
10^{-6}	micro	μ
10^{-9}	nano	n
10^{-12}	pico	p

VALUES OF TRIGONOMETRIC FUNCTIONS FOR COMMON ANGLES							
θ	$0°$	$30°$	$37°$	$45°$	$53°$	$60°$	$90°$
$\sin\theta$	0	1/2	3/5	$\sqrt{2}/2$	4/5	$\sqrt{3}/2$	1
$\cos\theta$	1	$\sqrt{3}/2$	4/5	$\sqrt{2}/2$	3/5	1/2	0
$\tan\theta$	0	$\sqrt{3}/3$	3/4	1	4/3	$\sqrt{3}$	∞

The following conventions are used in this exam.
  I.   The frame of reference of any problem is assumed to be inertial unless otherwise stated.
  II. In all situations, positive work is defined as work done on a system.
  III. The direction of current is conventional current: the direction in which positive charge would drift.
  IV. Assume all batteries and meters are ideal unless otherwise stated.
  V.   Assume edge effects for the electric field of a parallel plate capacitor unless otherwise stated.
  VI. For any isolated electrically charged object, the electric potential is defined as zero at infinite distance from the charged object.

ADVANCED PLACEMENT PHYSICS 2 EQUATIONS

MECHANICS

$$v_x = v_{x0} + a_x t$$

$$x = x_0 + v_{x0}t + \frac{1}{2}a_x t^2$$

$$v_x^2 = v_{x0}^2 + 2a_x(x - x_0)$$

$$\vec{a} = \frac{\sum \vec{F}}{m} = \frac{\vec{F}_{net}}{m}$$

$$\left|\vec{F}_f\right| \le \mu \left|\vec{F}_n\right|$$

$$a_c = \frac{v^2}{r}$$

$$\vec{p} = m\vec{v}$$

$$\Delta \vec{p} = \vec{F}\,\Delta t$$

$$K = \frac{1}{2}mv^2$$

$$\Delta E = W = F_{\parallel}d = Fd\cos\theta$$

$$P = \frac{\Delta E}{\Delta t}$$

$$\theta = \theta_0 + \omega_0 t + \frac{1}{2}\alpha t^2$$

$$\omega = \omega_0 + \alpha t$$

$$x = A\cos(\omega t) = A\cos(2\pi f t)$$

$$x_{cm} = \frac{\sum m_i x_i}{\sum m_i}$$

$$\vec{\alpha} = \frac{\sum \vec{\tau}}{I} = \frac{\vec{\tau}_{net}}{I}$$

$$\tau = r_{\perp}F = rF\sin\theta$$

$$L = I\omega$$

$$\Delta L = \tau\,\Delta t$$

$$K = \frac{1}{2}I\omega^2$$

$$\left|\vec{F}_s\right| = k|\vec{x}|$$

a	= acceleration
A	= amplitude
d	= distance
E	= energy
F	= force
f	= frequency
I	= rotational inertia
K	= kinetic energy
k	= spring constant
L	= angular momentum
ℓ	= length
m	= mass
P	= power
p	= momentum
r	= radius or separation
T	= period
t	= time
U	= potential energy
v	= speed
W	= work done on a system
x	= position
y	= height
α	= angular acceleration
μ	= coefficient of friction
θ	= angle
τ	= torque
ω	= angular speed

$$U_s = \frac{1}{2}kx^2$$

$$\Delta U_g = mg\,\Delta y$$

$$T = \frac{2\pi}{\omega} = \frac{1}{f}$$

$$T_s = 2\pi\sqrt{\frac{m}{k}}$$

$$T_p = 2\pi\sqrt{\frac{\ell}{g}}$$

$$\left|\vec{F}_g\right| = G\frac{m_1 m_2}{r^2}$$

$$\vec{g} = \frac{\vec{F}_g}{m}$$

$$U_G = -\frac{Gm_1 m_2}{r}$$

ELECTRICITY AND MAGNETISM

$$\left|\vec{F}_E\right| = \frac{1}{4\pi\varepsilon_0}\frac{|q_1 q_2|}{r^2}$$

$$\vec{E} = \frac{\vec{F}_E}{q}$$

$$\left|\vec{E}\right| = \frac{1}{4\pi\varepsilon_0}\frac{|q|}{r^2}$$

$$\Delta U_E = q\Delta V$$

$$V = \frac{1}{4\pi\varepsilon_0}\frac{q}{r}$$

$$\left|\vec{E}\right| = \left|\frac{\Delta V}{\Delta r}\right|$$

$$\Delta V = \frac{Q}{C}$$

$$C = \kappa\varepsilon_0\frac{A}{d}$$

$$E = \frac{Q}{\varepsilon_0 A}$$

$$U_C = \frac{1}{2}Q\Delta V = \frac{1}{2}C(\Delta V)^2$$

$$I = \frac{\Delta Q}{\Delta t}$$

$$R = \frac{\rho\ell}{A}$$

$$P = I\,\Delta V$$

$$I = \frac{\Delta V}{R}$$

$$R_s = \sum_i R_i$$

$$\frac{1}{R_p} = \sum_i \frac{1}{R_i}$$

$$C_p = \sum_i C_i$$

$$\frac{1}{C_s} = \sum_i \frac{1}{C_i}$$

$$B = \frac{\mu_0}{2\pi}\frac{I}{r}$$

A	= area
B	= magnetic field
C	= capacitance
d	= distance
E	= electric field
$\mathcal{E}$	= emf
F	= force
I	= current
ℓ	= length
P	= power
Q	= charge
q	= point charge
R	= resistance
r	= separation
t	= time
U	= potential (stored) energy
V	= electric potential
v	= speed
κ	= dielectric constant
ρ	= resistivity
θ	= angle
Φ	= flux

$$\vec{F}_M = q\vec{v}\times\vec{B}$$

$$\left|\vec{F}_M\right| = |q\vec{v}|\|\sin\theta\|\vec{B}|$$

$$\vec{F}_M = I\vec{\ell}\times\vec{B}$$

$$\left|\vec{F}_M\right| = |I\vec{\ell}|\|\sin\theta\|\vec{B}|$$

$$\Phi_B = \vec{B}\cdot\vec{A}$$

$$\Phi_B = |\vec{B}|\cos\theta|\vec{A}|$$

$$\mathcal{E} = -\frac{\Delta\Phi_B}{\Delta t}$$

$$\mathcal{E} = B\ell v$$

ADVANCED PLACEMENT PHYSICS 2 EQUATIONS

FLUID MECHANICS AND THERMAL PHYSICS

$$\rho = \frac{m}{V}$$

$$P = \frac{F}{A}$$

$$P = P_0 + \rho g h$$

$$F_b = \rho V g$$

$$A_1 v_1 = A_2 v_2$$

$$P_1 + \rho g y_1 + \frac{1}{2}\rho v_1{}^2$$

$$= P_2 + \rho g y_2 + \frac{1}{2}\rho v_2{}^2$$

$$\frac{Q}{\Delta t} = \frac{kA\,\Delta T}{L}$$

$$PV = nRT = Nk_B T$$

$$K = \frac{3}{2}k_B T$$

$$W = -P\Delta V$$

$$\Delta U = Q + W$$

A = area
F = force
h = depth
k = thermal conductivity
K = kinetic energy
L = thickness
m = mass
n = number of moles
N = number of molecules
P = pressure
Q = energy transferred to a system by heating
T = temperature
t = time
U = internal energy
V = volume
v = speed
W = work done on a system
y = height
ρ = density

WAVES AND OPTICS

$$\lambda = \frac{v}{f}$$

$$n = \frac{c}{v}$$

$$n_1 \sin\theta_1 = n_2 \sin\theta_2$$

$$\frac{1}{s_i} + \frac{1}{s_o} = \frac{1}{f}$$

$$|M| = \left|\frac{h_i}{h_o}\right| = \left|\frac{s_i}{s_o}\right|$$

$$\Delta L = m\lambda$$

$$d\sin\theta = m\lambda$$

d = separation
f = frequency or focal length
h = height
L = distance
M = magnification
m = an integer
n = index of refraction
s = distance
v = speed
λ = wavelength
θ = angle

GEOMETRY AND TRIGONOMETRY

Rectangle
$$A = bh$$

Triangle
$$A = \frac{1}{2}bh$$

Circle
$$A = \pi r^2$$
$$C = 2\pi r$$

Rectangular solid
$$V = \ell w h$$

Cylinder
$$V = \pi r^2 \ell$$
$$S = 2\pi r\ell + 2\pi r^2$$

Sphere
$$V = \frac{4}{3}\pi r^3$$
$$S = 4\pi r^2$$

A = area
C = circumference
V = volume
S = surface area
b = base
h = height
ℓ = length
w = width
r = radius

Right triangle
$$c^2 = a^2 + b^2$$

$$\sin\theta = \frac{a}{c}$$

$$\cos\theta = \frac{b}{c}$$

$$\tan\theta = \frac{a}{b}$$

MODERN PHYSICS

$$E = hf$$

$$K_{max} = hf - \phi$$

$$\lambda = \frac{h}{p}$$

$$E = mc^2$$

E = energy
f = frequency
K = kinetic energy
m = mass
p = momentum
λ = wavelength
ϕ = work function

AP PHYSICS 2

SECTION I

Time—90 minutes

50 Questions

Note: To simplify calculations, you may use $g = 10$ m/s^2 in all problems.

Directions: Each of the questions or incomplete statements below is followed by four suggested answers or completions. Select the one that is best in each case and mark it on your sheet.

1. What happens to the image formed by a concave mirror as the object is moved from far away to near the focal point?

    (A) The image moves away from the mirror and gets shorter.
    (B) The image moves away from the mirror and gets taller.
    (C) The image moves toward the mirror and gets shorter.
    (D) The image moves toward the mirror and gets taller.

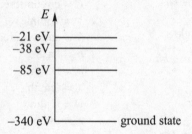

2. The diagram above (not drawn to scale) gives the first few electron energy levels within a single-electron atom. Which of the following gives the energy of a photon that could NOT be emitted by this atom during an electron transition?

    (A) 17 eV
    (B) 42 eV
    (C) 64 eV
    (D) 255 eV

3. During an isothermal expansion, a confined ideal gas does 150 J of work against its surroundings. Which of the following describes the heat transfer during this process?

    (A) 150 J of heat was added to the gas.
    (B) 150 J of heat was removed from the gas.
    (C) 300 J of heat was added to the gas.
    (D) 300 J of heat was removed from the gas.

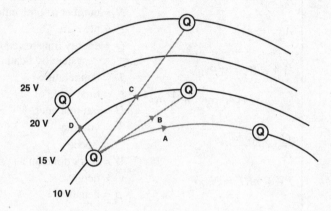

4. The figure above shows the equipotential lines a positive charge Q moves through. Along which path is the most work done on the charge?

    (A) Path A
    (B) Path B
    (C) Path C
    (D) Path D

5. An object of volume 2×10^{-3} m^3 and weight 6 N is placed into a tank of water, where it floats. What percentage of the object's volume is above the surface of the water?

    (A) 12%
    (B) 30%
    (C) 60%
    (D) 70%

GO ON TO THE NEXT PAGE.

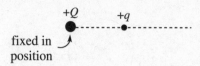

fixed in
position

6. The figure above shows two positively charged particles. The +Q charge is fixed in position, and the +q charge is brought close to +Q and released from rest. Which of the following graphs best depicts the acceleration of the +q charge as a function of its distance r from +Q ?

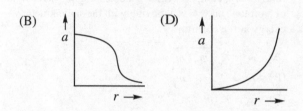

7. Each of the following particles is projected with the same speed into a uniform magnetic field **B** such that the particle's initial velocity is perpendicular to **B**. Which one would move in a circular path with the largest radius?

(A) Proton
(B) Beta particle
(C) Alpha particle
(D) Electron

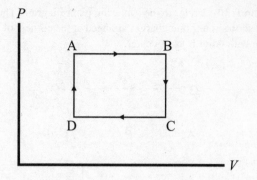

8. Which statement correctly characterizes the work done on the gas during the ABCA cycle shown in the above P–V diagram?

(A) There is no work done on the gas because the system both starts and concludes in state A.
(B) There is no work done because the work done during the step A→B cancels out the work done in step C→D.
(C) The work done on the gas is positive because the work done during the step A→B is greater than the work done in step C→D.
(D) The work done on the gas is positive because the work done during the step B→C is greater than the work done in step D→A.

9. How much current does a 60 W light bulb draw if it operates at a voltage of 120 V ?

(A) 0.25 A
(B) 0.5 A
(C) 2 A
(D) 4 A

GO ON TO THE NEXT PAGE.

Questions 10-11 refer to the following plane diagram. The figure shows four point charges arranged at the corners of a square with point B as its center.

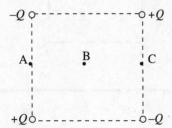

10. At which of the three points shown in the figure would a positive charge experience no net force?

    (A) A only
    (B) B only
    (C) C only
    (D) A, B, and C

11. At which points is the electric potential equal to zero?

    (A) A only
    (B) B only
    (C) C only
    (D) A, B, and C

12. A spherical balloon filled with helium is floating in air. If the balloon is inflated until its radius is doubled, how will the buoyant force on the balloon be affected?

    (A) It will decrease by a factor of 4.
    (B) It will increase by a factor of 4.
    (C) It will increase by a factor of 8.
    (D) It will not be affected.

13. A beam of monochromatic light entering a glass window pane from the air will experience a change in

    (A) frequency and wavelength
    (B) speed and wavelength
    (C) speed only
    (D) wavelength only

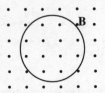

14. A metal loop of wire shown above is situated in a magnetic field **B** pointing out of the plane of the page. If **B** decreases uniformly in strength, the induced electric current within the loop is

    (A) clockwise and decreasing
    (B) clockwise and increasing
    (C) counterclockwise and decreasing
    (D) counterclockwise and constant

15. How does λ_e, the de Broglie wavelength of an electron (mass = m_e), compare with λ_p, the de Broglie wavelength of a proton (mass = m_p) moving with the same kinetic energy as the electron?

    (A) $\lambda_e = \lambda_p \sqrt{m_p / m_e}$
    (B) $\lambda_e = \lambda_p \sqrt{m_e / m_p}$
    (C) $\lambda_e = \lambda_p (m_p / m_e)^2$
    (D) $\lambda_e = \lambda_p (m_e / m_p)^2$

16. Data is collected in an experiment preformed on an ideal gas. In the experiment, temperature (in K) is the independent variable and volume (in m^3) is the dependent variable. If a plot is produced where the dependent variable is on the vertical axis, which of the following is true about the slope and y-intercept of the graph?

    (A) The slope will be linearly proportional to the pressure of the gas and the intercept will be 0 m^3.
    (B) The slope will be inversely proportional to the pressure of the gas and the intercept will be 0 m^3.
    (C) The slope will be linearly proportional to the pressure of the gas and the intercept will not be 0 m^3.
    (D) The slope will be inversely proportional to the pressure of the gas and the intercept will not be 0 m^3.

17. Which of the following changes to a double-slit interference experiment would increase the widths of the fringes in the interference pattern that appears on the screen?

    (A) Use light of a shorter wavelength.
    (B) Move the screen closer to the slits.
    (C) Move the slits closer together.
    (D) Use light with a lower wave speed.

GO ON TO THE NEXT PAGE.

$$^{13}_{6}C + (?) \rightarrow {}^{13}_{7}N + n$$

18. Identify the missing particle in the nuclear reaction given above.

    (A) Electron

    (B) Proton

    (C) Deuteron (a particle consisting of a proton and neutron)

    (D) Positron

Questions 19-20 refer to the circuit shown below:

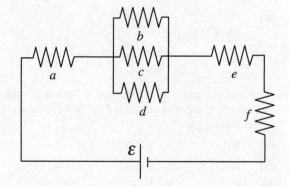

19. If each of the resistors in the circuit has a resistance of $3\,\Omega$, what must be the emf of the battery if the circuit draws a total current of 3 A ?

    (A) 0.3 V

    (B) 10 V

    (C) 30 V

    (D) 54 V

20. Which of the following relates P_b, the rate at which energy is dissipated by Resistor b, and P_e, the rate at which energy is dissipated by Resistor e ?

    (A) $P_b = 3P_e$

    (B) $P_b = 9P_e$

    (C) $P_e = 3P_b$

    (D) $P_e = 9P_b$

21. A capacitor C is connected to a resistor R in series. If the capacitor's initial charge was Q, what is the average power dissipated through the resistor if it loses half of its charge in a time t ?

    (A) $\dfrac{1}{8}\dfrac{Q^2}{Ct}$

    (B) $\dfrac{3}{8}\dfrac{Q^2}{Ct}$

    (C) $\dfrac{1}{8}\dfrac{Q^2C}{t}$

    (D) $\dfrac{3}{8}\dfrac{Q^2C}{t}$

22. If L, M, and T denote the dimensions of length, mass, and time, respectively, what are the dimensions of power?

    (A) L^2M/T^2

    (B) M^2L/T^2

    (C) L^2M/T^3

    (D) M^2L/T^3

23. Consider two adjacent transparent media. The speed of light in Medium 1 is v_1, and the speed of light in Medium 2 is v_2. If $v_1 < v_2$, then total internal reflection will occur at the interface between these media if a beam of light is

    (A) incident in Medium 1 and strikes the interface at an angle of incidence greater than $\sin^{-1}(v_1/v_2)$

    (B) incident in Medium 1 and strikes the interface at an angle of incidence greater than $\sin^{-1}(v_2/v_1)$

    (C) incident in Medium 2 and strikes the interface at an angle of incidence greater than $\sin^{-1}(v_1/v_2)$

    (D) incident in Medium 2 and strikes the interface at an angle of incidence greater than $\sin^{-1}(v_2/v_1)$

24. Traveling at an initial speed of 1.5×10^6 m/s, a proton enters a region of constant magnetic field of magnitude 1.5 T. If the proton's initial velocity vector makes an angle of 30° with the magnetic field, compute the proton's speed 4 s after entering the magnetic field.

    (A) 5.0×10^6 m/s

    (B) 7.5×10^6 m/s

    (C) 1.5×10^6 m/s

    (D) 3.0×10^6 m/s

25. A photocell of work function $\phi = 2eV$ is connected to a resistor in series. Light of frequency $f = 1 \times 10^{15}$ Hz hits a metal plate of the photocell. If the power of the light is $P = 100\,W$, what is the current through the resistor?

    (A) 2 A

    (B) 6 A

    (C) 12 A

    (D) 24 A

26. An electric dipole consists of a pair of equal but opposite point charges of magnitude 4.0 nC separated by a distance of 2.0 cm. What is the electric field strength at the point midway between the charges?

    (A) 0

    (B) 9.0×10^4 N/C

    (C) 1.8×10^5 N/C

    (D) 7.2×10^5 N/C

GO ON TO THE NEXT PAGE.

27. In an experiment designed to study the photoelectric effect, it is observed that low-intensity visible light of wavelength 550 nm produced no photoelectrons. Which of the following best describes what would occur if the intensity of this light were increased dramatically?

    (A) Almost immediately, photoelectrons would be produced with a kinetic energy equal to the energy of the incident photons.

    (B) Almost immediately, photoelectrons would be produced with a kinetic energy equal to the energy of the incident photons minus the work function of the metal.

    (C) After several seconds, necessary for the electrons to absorb sufficient energy from the incident light, photoelectrons would be produced with a kinetic energy equal to the energy of the incident photons minus the work function of the metal.

    (D) Nothing would happen.

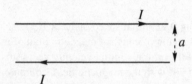

28. Electric current is established in each of two long parallel wires, as shown in the figure above. Describe the force per unit length on each wire.

    (A) Attractive, strength = $(\mu_0/2\pi)I^2/a^2$

    (B) Repulsive, strength = $(\mu_0/2\pi)I^2/a^2$

    (C) Attractive, strength = $(\mu_0/2\pi)I^2/a$

    (D) Repulsive, strength = $(\mu_0/2\pi)I^2/a$

29. An object is placed 100 cm from a plane mirror. How far is the image from the object?

    (A) 50 cm
    (B) 100 cm
    (C) 200 cm
    (D) 300 cm

30. An ideal fluid flows through a pipe with radius R and flow speed v. If the pipe splits up into three separate paths, each with radius $(R/2)$, what is the flow speed through each of the paths?

    (A) $1/3\,v$
    (B) $2/3\,v$
    (C) $4/3\,v$
    (D) $4v$

31. What happens to the pressure, P, of an ideal gas if the temperature is increased by a factor of 2 and the volume is increased by a factor of 8 ?

    (A) P decreases by a factor of 16.
    (B) P decreases by a factor of 4.
    (C) P decreases by a factor of 2.
    (D) P increases by a factor of 4.

32. Two nuclides that have the same excess number of neutrons over protons are called *isodiaphers*. Which of the following is an isodiaphere of $^{68}_{30}\text{Zn}$?

    (A) $^{67}_{31}\text{Ga}$
    (B) $^{74}_{32}\text{Ge}$
    (C) $^{72}_{34}\text{Se}$
    (D) $^{78}_{35}\text{Br}$

33. The plates of a capacitor are charged to a potential difference of 5 V. If the capacitance is 2 mF, what is the charge on the positive plate?

    (A) 0.005 C
    (B) 0.01 C
    (C) 0.02 C
    (D) 0.5 C

34. A confined ideal gas undergoes a cyclical process in three steps—an isobaric step, followed by an isochoric step, followed by an isothermal step. Which of the following must be true?

    (A) The change in internal energy of the gas is equal to the work done during the isobaric step.

    (B) The change in internal energy of the gas is equal to the work done during the isobaric step minus the work done during the isothermal step.

    (C) The total work done during the cycle is positive.

    (D) The total work done during the cycle is equal but opposite to the net amount of heat transferred.

35. A liquid flows at a constant flow rate through a pipe with circular cross-sections of varying diameters. At one point in the pipe, the diameter is 2 cm and the flow speed is 18 m/s. What is the flow speed at another point in this pipe, where the diameter is 3 cm?

    (A) 4 m/s
    (B) 6 m/s
    (C) 8 m/s
    (D) 12 m/s

GO ON TO THE NEXT PAGE.

36. A bi-convex lens has a radius of curvature of magnitude 20 cm. Which of the following best describes the image formed of an object of height 2 cm placed 30 cm from the lens?

    (A) Real, inverted, height = 1 cm
    (B) Virtual, upright, height = 0.25 cm
    (C) Real, upright, height = 1 cm
    (D) Virtual, inverted, height = 0.25 cm

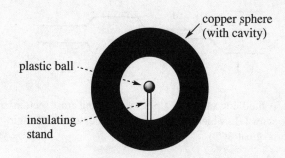

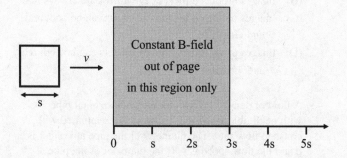

37. A square loop of conducting wire with side s is moved at a constant rate v to the right into a region where there is a constant magnetic field directed out of the page. Which of the following graphs shows the flux through the loop as a function of distance?

(A)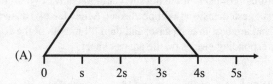

(B)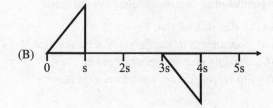

(C)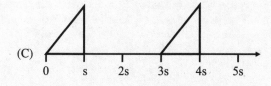

(D)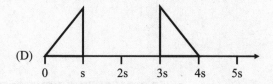

38. The figure above shows a solid copper sphere carrying a charge of +1 mC that contains a hollow cavity. Which of the following best describes what will occur when a plastic ball carrying a charge of +1 mC is placed inside the cavity?

    (A) A charge of +1 mC will appear on the inner wall of the cavity, leaving zero net charge on the outer surface of the sphere.
    (B) A charge of +1 mC will appear on the inner wall of the cavity, leaving a charge of –2 mC on the outer surface of the sphere.
    (C) A charge of –1 mC will appear on the inner wall of the cavity, leaving a charge of –2 mC on the outer surface of the sphere.
    (D) A charge of –1 mC will appear on the inner wall of the cavity, leaving a charge of +2 mC on the outer surface of the sphere.

39. Wire Y is made of the same material but has twice the diameter and half the length of Wire X. If Wire X has a resistance of R then Wire Y would have a resistance of

    (A) $R/8$
    (B) $R/2$
    (C) R
    (D) $2R$

40. How much work is required to charge a 10 μF capacitor to a potential difference of 100 V ?

    (A) 0.005 J
    (B) 0.01 J
    (C) 0.05 J
    (D) 0.1 J

GO ON TO THE NEXT PAGE.

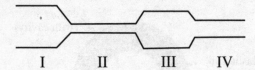

I   II   III   IV

41. A fluid flows through a pipe of changing cross sections as shown. In which section would the pressure of the fluid be greatest?

    (A) Section I
    (B) Section II
    (C) Section III
    (D) Section IV

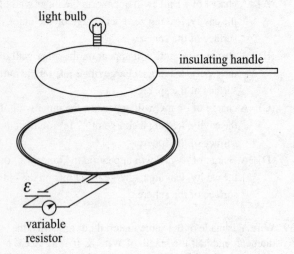

light bulb

insulating handle

ε

variable resistor

42. In the figure shown above, a loop of metal wire containing a tiny light bulb is attached to an insulating handle and placed over a coil of wire in which a current can be established by a source of emf and controlled by a variable resistor. The plane of the top loop is parallel to the plane of the bottom coil. Which of the following could NOT cause the bulb to light?

    (A) Rotating the handle 90° while keeping the plane of the top loop parallel to the plane of the bottom coil
    (B) Raising the handle up and away from the coil
    (C) Lowering the handle down toward the coil
    (D) Decreasing the resistance of the coil

43. Which of the following types of electromagnetic radiation has the longest wavelength?

    (A) Gamma rays
    (B) Ultraviolet
    (C) Blue light
    (D) Orange light

44. Which of the following statements about images is true?

    (A) Images produced by concave lenses are always real.
    (B) Images produced by concave mirros are always real.
    (C) Images produced by convex lenses can be both real and virtual.
    (D) Images produced by convex mirrors can be both real and virtual.

45. A fluid of density ρ flows through a horizontal pipe with negligible viscosity. The flow is streamlined with constant flow rate. The diameter of the pipe at Point 1 is d and the flow speed is v. If the diameter of the pipe at Point 2 is $d/3$, then the pressure at Point 2 is

    (A) less than the pressure at Point 1 by 4 ρv^2
    (B) less than the pressure at Point 1 by 40 ρv^2
    (C) more than the pressure at Point 1 by 4 ρv^2
    (D) more than the pressure at Point 1 by 40 ρv^2

Directions: For each of the questions 46-50 below, two of the suggested answers will be correct. Select the two answers that are best in each case, and then fill in both of the corresponding circles on the answer sheet.

46. Which of the following would increase the capacitance of a parallel-plate capacitor? Select two answers.

    (A) Using smaller plates
    (B) Inserting an insulating material between the plates
    (C) Moving the plates closer together
    (D) Increasing the voltage between the plates

GO ON TO THE NEXT PAGE.

47. You have a bar magnet and a coil connected to an ammeter. Which of the following actions would result in an induced current in the coil? Select two answers.

   (A) Putting the bar magnet close to the coil, but not moving it
   (B) Moving the bar magnet toward and away from the coil, perpendicular to the surface of the coil
   (C) Moving the bar magnet toward and away from the coil, along the surface of the coil
   (D) Placing the bar magnet at rest and rotating the coil near the bar magnet

48. The magnitude of an electric field in a vacuum at a point at a distance from a source charge is NOT dependent on which of the following? Select two answers.

   (A) Another charge placed at that point
   (B) The magnitude of the source charge
   (C) The distance from the source charge
   (D) The sign of the source charge

49. A metal conducting sphere of radius R in electrostatic equilibrium has a positive net charge of $+5e$. Which of the following is true? Select two answers.

   (A) All of the charge will be located on the outside of the sphere.
   (B) There will be a constant, non-zero electric field within the sphere.
   (C) There will be a constant, non-zero electric potential within the sphere.
   (D) When the sphere is connected to a ground, the $5e$ of charge on the sphere will flow into the ground to neutralize the sphere.

50. A circuit is created with a battery of negligible internal resistance and three identical resistors with resistance of $10\ \Omega$. The resistors are originally arranged so that one is in series with the battery, and the other two are in parallel with one another. Which of the following changes to the circuit will result in an increase in the amount of current drawn from the battery? Select two answers.

   (A) Rearranging the resistors so that all three are in parallel
   (B) Replacing every resistor with a resistor of half the resistance
   (C) Removing one of the two parallel branches entirely from the circuit
   (D) Replacing the battery with a battery with half the voltage

END OF SECTION I

AP PHYSICS 2

SECTION II

Time—90 minutes

4 Questions

Directions: Questions 1 and 2 are long free-response questions that require about 30 minutes to answer. Questions 3 and 4 are short answer questions that require about 15 minutes to answer. Show your work for each part in the space provided after that part.

1. The figure below shows an electric circuit containing a source of emf, $\mathcal{E}$, a variable resistor (r) and a resistor of fixed resistance R. The resistor R is immersed in a sealed beaker containing a mass m of water, currently at temperature T_i. When the switch S is closed, current through the circuit causes the resistor in the water to dissipate heat, which is absorbed by the water. A stirrer at the bottom of the beaker simply ensures that the temperature is uniform throughout the water at any given moment. The apparatus is well-insulated (insulation not shown), and it may be assumed that no heat is lost to the walls or lid of the beaker or to the stirrer.

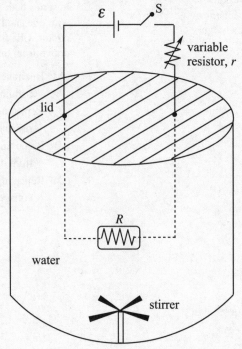

(a) Determine the current in the circuit once S is closed. Write your answer in terms of $\mathcal{E}$, r, and R.

(b) How much heat is dissipated by the resistor R in time t?

(c) Explain briefly how the temperature of the water can be increased more rapidly by adjusting the rotation rate of the stirrer.

(d) As the temperature of the water increases, explain the microscopic interactions responsible for the changing pressure in the container.

GO ON TO THE NEXT PAGE.

2. A *cyclotron* is a device used to accelerate charged particles to high speeds. It consists of two hollow containers—called *dees* because of their shape—facing each other and separated by a small gap. They are immersed in a uniform magnetic field, **B**, and are attached to a source of alternating voltage. A charged particle is projected from the center of the cyclotron into Dee #1, and the magnetic force causes it to turn in a circle. When it completes its semicircular path within one dee, the polarity of the voltage is reversed, and the particle is accelerated across the gap into the adjacent dee. This process continues, and the particle spirals outward at faster and faster speeds, until it emerges from Dee #2. Notice that the voltage must be alternated twice during each revolution of the particle. The figure below shows a view—looking down from above—of a cyclotron.

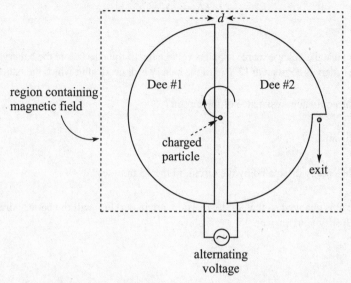

(a) Explain why the electric field in the gap must be used to increase the speed of the particle.

Assume that the charged particle to be accelerated by the cyclotron is a proton.

(b) Should the magnetic field point into the plane of the page or out of the plane of the page in order to cause the particle to rotate clockwise as shown in the figure?

For each of the following, write your answer in terms of the magnetic field strength B, the proton's charge (+e) and mass (m), and fundamental constants.

(c) Show that the time to complete one revolution does not depend on the speed of the proton, and determine this orbital period.

(d) How many revolutions does the proton make per second?

(e) What must be the frequency (in Hz) of the alternating voltage?

(f) If the maximum radius of the proton's orbit is R, what is its maximum kinetic energy upon exiting? (Your answer should also include R.)

GO ON TO THE NEXT PAGE.

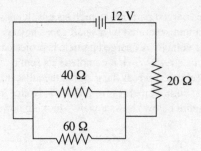

3. The circuit shown is built so that the voltage source supplies voltage for 15 minutes before the battery is completely drained. Assume the voltage supplied by the battery is constant at 12 V until the battery is drained, after which the battery supplies 0 V.

    (a) What is the equivalent resistance of the circuit?

    (b) What is the power?

    (c) What is the energy dissipated by the circuit in the 15 minutes?

    (d) If the circuit is changed so that all the resistors are in parallel, will the battery drain more quickly or less quickly? Justify your answer.

GO ON TO THE NEXT PAGE.

4. A cylindrical tank 50 m high and 60 m in diameter resting on level ground is completely filled with water which is open to the air. A worker accidentally pokes a small hole in the side of the tank, 35 m above the base. The water then flows out through the hole at a rate of $3 \times 10^{-4}\,\text{m}^3/\text{min}$. The tank is shown in the figure below.

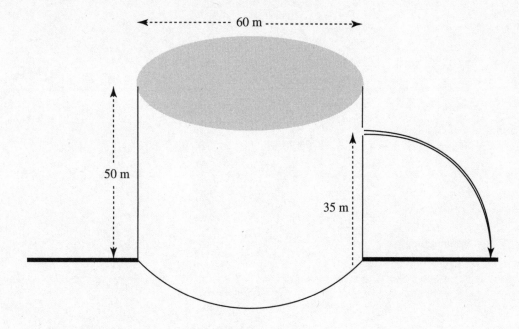

(a) Determine the speed with which the water exits the hole.

(b) What is the diameter of the hole?

(c) How far from the base of the tank does the stream of water initially hit the ground?

STOP

END OF EXAM

Practice Test 1:
Answers and
Explanations

PRACTICE TEST 1 ANSWER KEY

1.	B	26.	D
2.	B	27.	D
3.	A	28.	D
4.	C	29.	C
5.	D	30.	C
6.	A	31.	B
7.	C	32.	D
8.	C	33.	B
9.	B	34.	D
10.	B	35.	C
11.	D	36.	A
12.	C	37.	A
13.	B	38.	D
14.	D	39.	A
15.	A	40.	C
16.	B	41.	A
17.	C	42.	A
18.	B	43.	D
19.	C	44.	C
20.	D	45.	B
21.	B	46.	B, C
22.	C	47.	B, D
23.	A	48.	A, D
24.	C	49.	A, C
25.	D	50.	A, B

SECTION I

1. **B** One way to answer this question is to invent some numbers that satisfy the conditions of the question and observe what happens. Let's choose $f = 1$ m, and take $s_{o1} = 3$ m and $s_{o2} = 2$ m (notice that $f < s_{o2} < s_{o1}$, which means that the object is moving closer to the focal point). Using the mirror equation, we find that

$$\frac{1}{s_{o1}} + \frac{1}{s_{i1}} = \frac{1}{f} \quad \Rightarrow \quad \frac{1}{3 \text{ m}} + \frac{1}{s_{i1}} = \frac{1}{1 \text{ m}} \quad \Rightarrow \quad s_{i1} = \frac{3}{2} \text{ m}$$

$$\frac{1}{s_{o2}} + \frac{1}{s_{i2}} = \frac{1}{f} \quad \Rightarrow \quad \frac{1}{2 \text{ m}} + \frac{1}{s_{i2}} = \frac{1}{1 \text{ m}} \quad \Rightarrow \quad s_{i2} = 2 \text{ m}$$

Since $s_{i2} > s_{i1}$, the image is moving away from the mirror (eliminating (C) and (D)). Now for the magnifications:

$$M_1 = -\frac{s_{i1}}{s_{o1}} = -\frac{\frac{3}{2} \text{ m}}{3 \text{ m}} = -\frac{1}{2}$$

$$M_2 = -\frac{s_{i2}}{s_{o2}} = -\frac{2 \text{ m}}{2 \text{ m}} = -1$$

Since $|M_2| > |M_1|$, the image gets taller. (You can also use ray diagrams.)

2. **B** The energies emitted during electron transitions are equal to the differences between the allowed energy levels. Choice (A), 17 eV, is equal to the energy emitted when the electron drops from the −21 eV level to the −38 eV level. Choice (C), 64 eV, is equal to the energy emitted when the electron drops from the −21 eV level to the −85 eV level. Choice (D), 255 eV, is equal to the energy emitted when the electron drops from the −85 eV level to the −340 eV level. However, no electron transition in this atom could give rise to a 42 eV photon.

3. **A** Because the expansion is isothermal, no change in temperature occurs, which implies that no change in the internal energy of the gas takes place. Since $\Delta U = Q + W$ (the first law of thermodynamics), the fact that $\Delta U = 0$ implies that $Q = -W$. Since $W = -150$ J, it must be true that $Q = +150$ J.

4. **C** We know that $W_E = -\Delta U_E$ and that $\Delta U_E = q\Delta V$. So, the largest work done is the path that has the largest change in potential, which is path C. Furthermore, since this is a positive charge (which wants to move to a lower potential), we know work is done on it to move it to a higher potential.

5. **D** The mass of the object is $m = w/g = (6 \text{ N})/(1 \text{ kg}/10 \text{ N}) = 0.6$ kg, so the density of the object is $\rho = m/V = (0.6 \text{ kg})/(2 \times 10^{-3} \text{ m}^3) = 300$ kg/m³. Since this is 30% of the density of water, 30% of the floating object's volume will be submerged. This leaves 70% of the object's volume above the surface of the water.

6. **A** If the mass of the $+q$ charge is m, then its acceleration is

$$a = \frac{F_E}{m} = \frac{1}{4\pi\varepsilon_0} \frac{Qq}{mr^2}$$

The graph in (A) best depicts an inverse-square relationship between a and r.

7. **C** When the particle enters the magnetic field, the magnetic force provides the centripetal force to cause the particle to execute uniform circular motion:

$$|q|vB = \frac{mv^2}{r} \quad \Rightarrow \quad r = \frac{mv}{|q|B}$$

Since v and B are the same for all the particles, the largest r is found by maximizing the ratio $m/|q|$. The value of $m/|q|$ for an alpha particle is about twice that for a proton and thousands of times greater than that of an electron or positron.

8. **C** Work is done when there is a change in volume. At constant pressure, the equation $W = -P\Delta V$ tells us that when P is higher (as it is at path A $\rightarrow$ B) a greater amount of negative work is done on the gas than at lower P. Thus, the work done by the gas during the entire cycle is positive, and the constant volume paths have no influence on the amount of work.

9. **B** We use the equation for the power dissipated by a resistor, $P = IV$:

$$I = \frac{P}{V} = \frac{60 \text{ W}}{120 \text{ V}} = 0.5 \text{ A}$$

10. **B** A point charge placed at point A, B, or C would experience no net force only if the electric field at that point were zero. Let's label the four charges 1, 2, 3, and 4, starting at the upper right and counting counterclockwise. The diagram below shows the individual electric field vectors at points A, B, and C due to the four charges.

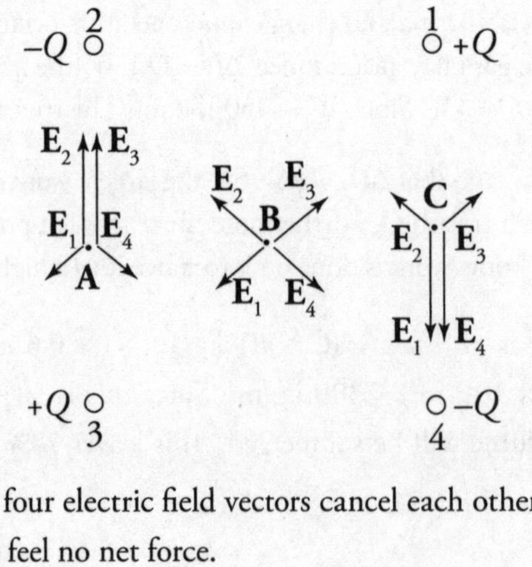

Only at point B do the four electric field vectors cancel each other to give a sum of zero, so only at point B would a charge feel no net force.

11. **D** The electric potential is a scalar, given by the formula $\phi = kQ/r$, where $k = 1/4\pi\varepsilon_0$ is Coulomb's constant and r is the distance from the charge Q. In this case, the total potential at any point is the sum of the four potentials produced by the four source charges. Here's the computation of the potential at each of the points A, B, and C, where a denotes the length of each side of the square in the diagram. Note that all of them equal zero.

$$\phi_A = \phi_{A1} + \phi_{A2} + \phi_{A3} + \phi_{A4} = k\frac{+Q}{\frac{\sqrt{5}}{2}a} + k\frac{-Q}{\frac{1}{2}a} + k\frac{+Q}{\frac{1}{2}a} + k\frac{-Q}{\frac{\sqrt{5}}{2}a} = 0$$

$$\phi_B = \phi_{B1} + \phi_{B2} + \phi_{B3} + \phi_{B4} = k\frac{+Q}{\frac{\sqrt{2}}{2}a} + k\frac{-Q}{\frac{\sqrt{2}}{2}a} + k\frac{+Q}{\frac{\sqrt{2}}{2}a} + k\frac{-Q}{\frac{\sqrt{2}}{2}a} = 0$$

$$\phi_C = \phi_{C1} + \phi_{C2} + \phi_{C3} + \phi_{C4} = k\frac{+Q}{\frac{1}{2}a} + k\frac{-Q}{\frac{\sqrt{5}}{2}a} + k\frac{+Q}{\frac{\sqrt{5}}{2}a} + k\frac{-Q}{\frac{1}{2}a} = 0$$

12. **C** The buoyant force is given by $F_B = \rho_{air}V_{sub}g$. Because the entire balloon is surrounded by air, V_{sub} is equal to V, the full volume of the balloon. Therefore, the buoyant force is proportional to the balloon's volume. Since the volume of a sphere is given by $V = \frac{4}{3}\pi r^3$, if r is doubled, V will increase by a factor of $2^3 = 8$. Thus, F_B increases by a factor of 8.

13. **B** When a wave enters a new medium, its frequency does not change, but its wave speed does (Wave Rule 2). Since $\lambda f = v$, the change in wave speed implies a change in wavelength also.

14. **D** Since the magnetic flux out of the page is decreasing, the induced current will oppose this change (as always), attempting to create more magnetic flux out of the page. In order to do this, the current must circulate in the counterclockwise direction (remember the right-hand rule). As B decreases uniformly (that is, while $\Delta B/\Delta t$ is negative and constant), the induced emf,

$$\varepsilon = -\frac{\Delta\Phi_B}{\Delta t} = -A\frac{\Delta B}{\Delta t}$$

is nonzero and constant, which implies that the induced current, $I = \varepsilon/R$, is also nonzero and constant.

15. **A** The de Broglie wavelength of a particle moving with linear momentum of magnitude $p = mv$ is given by the equation $\lambda = h/p$, where h is Planck's constant. Because

$$p = mv = \sqrt{m(mv^2)} = \sqrt{2m\left(\frac{1}{2}mv^2\right)} = \sqrt{2mK}$$

we find that

$$\frac{\lambda_e}{\lambda_p} = \frac{h/\sqrt{2m_eK}}{h/\sqrt{2m_pK}} = \frac{\sqrt{2m_pK}}{\sqrt{2m_eK}} = \sqrt{\frac{m_p}{m_e}} \implies \lambda_e = \lambda_p\sqrt{\frac{m_p}{m_e}}$$

16. **B**   The ideal gas law equation is $PV = nRT$. Solving this for V (since volume is on the vertical axis of the graph described in the problem) yields $V = (nR/P)T$. Comparing this to the equation for a line, $y = mx + b$, the slope of the line will be (nR/P), which is inversely proportional to P, and $b = 0$ is the intercept.

17. **C**   Relative to the central maximum, the locations of the bright fringes on the screen are given by the expression $m(\lambda L/d)$, where λ is the wavelength of the light used, L is the distance to the screen, d is the separation of the slits, and m is an integer. The width of a fringe is therefore $(m + 1)(\lambda L/d) - m(\lambda L/d) = \lambda L/d$. One way to increase $\lambda L/d$ is to decrease d.

18. **B**   Nuclear reactions must conserve mass number and charge, so the missing particle must be a proton:

$$^{13}_{6}\text{C} + {}^{1}_{1}\text{p} \rightarrow {}^{13}_{7}\text{N} + {}^{1}_{0}\text{n}$$

19. **C**   Resistors a, b, and c may be summed together as parallel resistors to find their equivalent resistance.

$$\left(\left(\frac{1}{3\,\Omega}\right) + \left(\frac{1}{3\,\Omega}\right) + \left(\frac{1}{3\,\Omega}\right)\right)^{-1} = \frac{1}{1\,\Omega}$$

Then, the equivalent resistance of 3 ohms in parallel with resistors a, e, and f are added together in series. $(3\,\Omega + 1\,\Omega + 3\,\Omega + 3\,\Omega) = 10\,\Omega$. $V = IR$, and thus we find that $V = (3\text{ A})(10\,\Omega) = 30\text{ V}$.

20. **D**   The power dissipated by a resistor is I^2R. Since the current through Resistor b is $\frac{1}{3}$ of the current through Resistor e, we find that

$$\frac{P_b}{P_e} = \frac{I_b^2 R}{I_e^2 R} = \frac{I_b^2}{I_e^2} = \frac{\left(\frac{1}{3}I_e\right)^2}{I_e^2} = \frac{1}{9} \quad \Rightarrow \quad P_e = 9P_b$$

21. **B**   We know that $U_C = \frac{1}{2}CV^2$ and that $V = Q/C$. Combining these, we get $U_C = \frac{1}{2}\frac{Q^2}{C}$. Now, the initial charge is Q and the final charge is $\frac{Q}{2}$, so the change in potential energy is

$$|\Delta U_C| = \frac{1}{2}\frac{Q^2}{C} - \frac{1}{2}\frac{(Q/2)^2}{C} = \frac{1}{2}\frac{Q^2}{C} - \frac{1}{8}\frac{Q^2}{C} = \left(\frac{4}{8} - \frac{1}{8}\right)\frac{Q^2}{C} = \frac{3}{8}\frac{Q^2}{C}$$

So, the average power dissapated by the resistor is $P_{avg} = |\Delta U_C|/t$, or

$$P_{avg} = \frac{3}{8}\frac{Q^2}{Ct}$$

22. **C**   The SI unit for power, P, is the watt, where 1 watt = 1 joule per second. One joule is equal to 1 newton-meter, and 1 newton = 1 kg·m/s². Therefore,

$$[P] = \frac{\text{J}}{\text{s}} = \frac{\text{N}\cdot\text{m}}{\text{s}} = \frac{\frac{\text{kg}\cdot\text{m}}{\text{s}^2}\cdot\text{m}}{\text{s}} = \frac{\text{kg}\cdot\text{m}^2}{\text{s}^3} \quad \Rightarrow \quad [P] = \frac{\text{L}^2\text{M}}{\text{T}^3}$$

23. **A**   In order for total internal reflection to occur, the beam must be incident in the medium with the higher index of refraction and strike the interface at an angle of incidence greater than the critical angle. Since $v_1 < v_2$, the refractive index of Medium 1, $n_1 = c/v_1$, must be *greater* than the index of Medium 2 ($n_2 = c/v_2$), and the critical angle is $\theta_c = \sin^{-1}(n_2/n_1) = \sin^{-1}(v_1/v_2)$.

24. **C**   Since the magnetic force is always perpendicular to the object's velocity, it does zero work on any charged particle. Zero work means zero change in kinetic energy, so the speed remains the same. Remember: The magnetic force can only change the direction of a charged particle's velocity, not its speed.

25. **D**   The energy of the incoming light is $E = hf = (4.14 \times 10^{-15}\,\text{eVs})(1 \times 10^{15}\,\text{Hz}) = 4.14$ eV. This means that a photocurrent can be produced, since the energy of the incoming photon is greater than the work function of the metal plate. Now, this energy in joules is $E = (4.14\ \text{eV})(1.6 \times 10^{-19}\,\text{J/eV}) = 6.626 \times 10^{-19}$ J, so the number of photons emitted per second is

$$n = \frac{P}{E} = \frac{100\,W}{6.626 \times 10^{-19}} = 1.51 \times 10^{20}\,\text{s}^{-1}$$

This means that there are 1.51×10^{20} electrons ejected from the plate per second, so the current is $I = ne = (1.51 \times 10^{20}\,\text{s}^{-1})(1.6 \times 10^{-19}\,\text{C}) = 24$ A

26. **D**   By definition, an electric dipole consists of a pair of charges that are equal in magnitude but opposite in sign. The electric field at the point midway between the charges is equal to the sum of the electric fields due to each charge alone:

Because the individual electric field vectors point in the same direction, their magnitudes add to give the total electric field:

$$E_{\text{total}} = 2E = 2\frac{kQ}{r^2} = 2\frac{kQ}{\left(\frac{1}{2}d\right)^2} = 8\frac{kQ}{d^2} = 8 \cdot \frac{(9 \times 10^9\ \text{N} \cdot \text{m}^2/\text{C}^2)(4.0 \times 10^{-9}\ \text{C})}{(2.0 \times 10^{-2}\ \text{m})^2}$$

$$= 7.2 \times 10^5\ \text{N/C}$$

27. **D**   If the photons of the incident light have insufficient energy to liberate electrons from the metal's surface, then simply increasing the number of these weak photons (that is, increasing the intensity of the light) will do nothing. In order to produce photoelectrons, each photon of the incident light must have an energy at least as great as the work function of the metal.

28. **D**   The force per unit length on each wire is $(\mu_0/2\pi)I_1 I_2/r$, where r is the separation distance. Since $I_1 = I_2 = I$ and $r = a$, the force per unit length on each wire is $(\mu_0/2\pi)I^2/a$. Remember that when the currents are in opposite directions, the force between the wires is repulsive.

29. **C** The distance from the mirror to the image is equal to the distance from the mirror to the object. Therefore, the distance from the object to the image is 100 cm + 100 cm = 200 cm.

30. **C** If flow rate is constant for ideal fluids, then it becomes an inverse relationship between v, flow speed, and A, the cross-section area of the pipe

$$A_1 v_1 = A_2 v_2 \rightarrow \pi R^2 v_f = \pi (R/2)^2 v \rightarrow v_f = 4v$$

but there are 3 pipes, so $4/3\ v$.

31. **B** Use the Ideal Gas Law:

$$P' = \frac{nRT'}{V'} = \frac{nR(2T)}{8V} = \frac{1}{4}\frac{nRT}{V} = \frac{1}{4}P$$

32. **D** The nuclide $^{68}_{30}\text{Zn}$ has 30 protons and 68 − 30 = 38 neutrons, so the excess number of neutrons over protons is 38 − 30 = 8. Of the choices given, only the nuclide $^{78}_{35}\text{Br}$ shares this property; it has 35 protons and 78 − 35 = 43 neutrons, giving an excess of neutrons over protons of 43 − 35 = 8.

33. **B** The charge on each plate has magnitude $Q = CV = (2 \times 10^{-3}\ \text{F})(5\ \text{V}) = 0.01$ C.

34. **D** Since the process is cyclic, $\Delta U = 0$, neither choice (A) nor (B) can be true. Choice (C) is not true; whether a cycle is clockwise or counterclockwise in a P–V diagram will determine the sign of the total work. But choice (D) must be true; since $\Delta U = 0$, the First Law of Thermodynamics says that $W = -Q$.

35. **C** The Continuity Equation, Av = constant, tells us that v, the flow speed, is inversely proportional to A, the cross-sectional area of the pipe. Since A is proportional to d^2, where d is the diameter of the pipe, we can say that v is inversely proportional to d^2. Now, if d increases by a factor of 3/2 (from 2 cm to 3 cm), then v decreases by a factor of $(3/2)^2 = 9/4$. Thus, the flow speed at the point where the pipe is 3 cm in diameter is $(4/9)(18\ \text{m/s}) = 8$ m/s.

36. **A** A bi-convex lens is a converging lens, so it has a positive focal length. Since $f = R/2$, we have $f = (20\ \text{cm})/2 = 10$ cm. The lens equation then gives

$$\frac{1}{s_o} + \frac{1}{s_i} = \frac{1}{f} \quad \Rightarrow \quad \frac{1}{30\ \text{cm}} + \frac{1}{s_i} = \frac{1}{10\ \text{cm}} \quad \Rightarrow \quad s_i = 15\ \text{cm}$$

Because s_i is positive, the image is real, and real images are inverted.

This is enough information to eliminate (B), (C), and (D). We can also just calculate the magnification:

$$m = \frac{s_i}{s_o} = \frac{15\ \text{cm}}{30\ \text{cm}} = 0.5$$

So, the image is 0.5 times the height of the object, or 1 cm.

37.   **A**   The flux depends on the field strength and the area of the loop experiencing the field. The flux linearly increases as the loop moves into the region of the magnetic field. Once the loop is fully within the magnetic field region, the flux remains constant until the loop begins to exit the magnetic field region. As the loop begins to leave the region of the magnetic field, the flux decreases until it is 0 again once the loop is completely out of the region of the field. This is consistent with (A).

38.   **D**   Because no electrostatic field can exist within the body of a conductor, a charge of –1 mC will appear on the wall of the cavity to "guard" the +1 mC charge of the plastic ball. Since the sphere has a charge of +1 mC, a net charge of (+1 mC) – (–1 mC) = +2 mC will appear on the outer surface of the copper sphere.

39.   **A**   The resistance of a wire is given by the formula $R = \rho L / A$, where ρ is the resistivity of the wire, L is its length, and A is its cross-sectional area. Since both wires are made of the same material, the value of ρ is the same for both wires, so we only need to look at the ratio L/A. Because the diameter of Wire Y is twice the diameter of Wire X, the cross-sectional area of Wire Y is 4 times the cross-sectional area of Wire X. Since

$$\frac{L_Y}{A_Y} = \frac{\frac{1}{2}L_X}{4A_X} = \frac{1}{8}\frac{L_X}{A_X}$$

we see that the resistance of wire Y is 1/8 the resistance of wire X.

40.   **C**   The work required to charge the capacitor is equal to the resulting increase in the stored electrical potential energy:
$$U = \frac{1}{2}CV^2 = \frac{1}{2}(10 \times 10^{-6} \text{ F})(100 \text{ V})^2 = 0.05 \text{ J}$$

41.   **A**   According to the Bernoulli effect, the flow speed will be lowest, and thus the pressure will be greatest, in the section of the pipe with the greatest diameter.

42.   **A**   Choices (B), (C), and (D) all change the magnetic flux through the loop containing the light bulb, thus inducing an emf and a current. However, just swinging the handle over as described in choice A changes neither the area presented to the magnetic field lines from the bottom coil nor the density of the field lines at the position of the loop. No change in magnetic flux means no induced emf and no induced current.

43.   **D**   Gamma rays and X-rays are very high-energy, short-wavelength radiations. Ultraviolet light has a higher energy and shorter wavelength than the visible spectrum. Within the visible spectrum, the colors are—in order of increasing frequency—ROYGBV, so orange light has a lower frequency (and thus longer wavelength) than blue light.

44.   **C**   The equation needed to address this question is

$$\frac{1}{s_i} = \frac{1}{f} - \frac{1}{s_o}$$

Convex lenses and concave mirrors have positive focal lengths, and concave lenses and convex mirrors have negative focal lengths. If the focal length is negative, then s_i is always negative. This eliminates (A) and (D). If the focal length is positive, then s_i could be either positive or negative. This eliminates (B), leaving (C) as the only correct answer.

45.   **B**   Speed becomes greater when the cross-sectional Area becomes smaller, so pressure is lower. Eliminate (C) and (D). Since the diameter has decreased by 3, the cross-sectional area decreases by a factor of 9. This means the flow speed must increase by a factor of 9.

$$v_2 = 9v$$

Using Bernoulli's Equation

$$P_1 + \frac{1}{2}\rho v_1^2 + \rho g y_1 = P_2 + \frac{1}{2}\rho v_2^2 + \rho g y_2$$

Because there is no change in height we can eliminate it from both sides of the equation

$$P_1 + \frac{1}{2}\rho v_1^2 = P_2 + \frac{1}{2}\rho v_2^2$$
$$P_1 - P_2 = \frac{1}{2}\rho (v_2^2 - v_1^2) = \frac{1}{2}\rho ((9v)^2 - v^2) = 40\rho v^2$$

46.   **B, C**   Placing a dielectric (insulator) in between the plates of a capacitor always increases the capacitance. Choice (B) will increase the capacitance. A capacitor's capacitance is determined by the area of its plates, A, and the distance of separation, d, between the plates as given in the equation $C = \varepsilon_0 A/d$. By moving the plates closer we increase the capacitance. Choice (C) will also increase the capacitance.

47.   **B, D**   To induce a current in a coil, you need to change the magnetic flux through the coil. The magnetic flux depends upon three things: the magnitude of the magnetic field, the area of the coil, and the angle that the magnetic field makes to the line perpendicular to the surface of the coil. In (A), the situation is static and nothing is changing, so no current can be produced. In (C), the bar magnet is moving (which would change the magnitude of the magnetic field), but the direction of the magnetic field is along the surface of the coil. This means that the component of the magnetic field perpendicular to the surface of the coil is zero and therefore the flux is always zero. This leaves (B) (a change in magnitude of magnetic field) and (D) (a change in the angle) as the actions that would produce a current in the coil.

48. **A, D** The strength of the electric field due to a single point charge is given by the equation $E = kQ/r^2$. Thus, E clearly depends on both the magnitude of the source charge Q and the distance r from Q; this eliminates (B) and (C). The sign of the source charge affects only the direction of the electric field vectors; they will point away from the source charge if it is positive and toward the source charge if it is negative. The sign of the source charge does not affect the strength of the field. Charges do not create an electric field at the location of the charge itself. Therefore, the electric field at that same point is not changed by the addition of a charge at that point.

49. **A, C** Conductors will always have all of their excess charge located at the surface. The entire volume of a conductor must be an equipotential in electrostatics because otherwise the charges in the conductor would move as a result of the potential difference. Choice (B) is false because within a conductor in electrostatic equilibrium, there is an electric field of 0 N/C within the conductor. Choice (D) is false because, while the sphere will neutralize when connected to a ground, it is not the positive charges that flow out of the sphere but negative charges which will flow from the ground into the sphere.

50. **A, B** From Ohm's Law, the current drawn out of the battery will increase when the total resistance of the circuit decreases, or when the total voltage supplied to the circuit increases, which makes (D) false. Resistors in parallel have a lower equivalent resistance than those in series, so (A) is true. To see that (B) decreases the overall resistance, we can think of a simpler version of the problem. Replacing a resistor with a wire will have the same effect of either increasing or decreasing the overall resistance, but with a larger effect. If any of the three resistors are replaced with a wire, the overall resistance decreases. Therefore, decreasing the resistance of any of the resistors will decrease the equivalent resistance, and replacing all of them most certainly will decrease the equivalent resistance.

SECTION II

1. (a) Since the total resistance in the circuit is $r + R$, Ohm's law gives

$$I = \frac{\mathcal{E}}{r + R}$$

(b) The power dissipated as heat by resistor R is $P = I^2R$; multiplying this by the time, t, gives Q, the heat dissipated:

$$Q = Pt = I^2Rt = \frac{\mathcal{E}^2 Rt}{(r + R)^2}$$

(c) The movement of the blades of the stirrer through the water transfers mechanical work to thermal energy. The faster the stirrer rotates (i.e., the more rotational kinetic energy it has), the more heat will be transferred to the water and the faster the temperature will rise. [In fact, given enough time, the rotating stirrer could bring the water to boil all by itself (assuming that heat losses to the beaker, the lid, etc. could be kept low)].

(d) As the temperature increases, the speed of the water molecules increases also. Therefore, the water molecules will strike the edges of the container more frequently. As a result of the increase in collision between the boundary and the water molecules, the pressure in the container will increase.

2. (a) Magnetic forces alone can only change the direction of a charged particle's velocity, not its speed. This is because $\mathbf{F}_B = q(\mathbf{v} \times \mathbf{B})$ is always perpendicular to $\mathbf{v}$, the particle's velocity, so it can do no work and thus cause no change in kinetic energy. The electric field is required to increase the particle's speed.

(b) At the moment shown below, $\mathbf{v}$ is upward in the plane of the page, and $\mathbf{F}$ must point toward the center of the circular path.

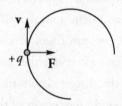

In order for $\mathbf{F}_B = q(\mathbf{v} \times \mathbf{B})$ to be consistent with this, the right-hand rule implies that $\mathbf{B}$ must point out of the plane of the page (because q is positive).

(c) The time to complete one revolution, T, is equal to $2\pi r/v$, where r is the radius of the path and v is the speed. Since the magnetic force provides the centripetal force,

$$qvB = \frac{mv^2}{r} \;\Rightarrow\; qB = \frac{mv}{r} \;\Rightarrow\; \frac{r}{v} = \frac{m}{qB} \;\Rightarrow\; T = \frac{2\pi r}{v} = \frac{2\pi m}{qB} = \frac{2\pi m}{eB}$$

Therefore, T depends only on e, m, and B.

(d) Since T is the time for one revolution, the number of revolutions per unit time is the reciprocal of T, $eB/(2\pi m)$.

(e) The description given with the problem states that "the voltage must be alternated twice during each revolution of the particle." Therefore, the frequency of the alternating voltage must be twice the value found in part (d), $2eB/(2\pi m) = eB/(\pi m)$.

(f) From part (c), we know that

$$evB = \frac{mv^2}{r} \Rightarrow v = \frac{eBr}{m}$$

Therefore,

$$K_{max} = \frac{1}{2}mv_{max}^2 = \frac{1}{2}m\left(\frac{eBr_{max}}{m}\right)^2 = \frac{1}{2}m\left(\frac{eBR}{m}\right)^2 = \frac{e^2B^2R^2}{2m}$$

3. (a) The equivalent resistance of the circuit can be found by first calculating the resistance of the two resistors in parallel and then adding that value to the resistor in series. For parallel circuits:

$$\frac{1}{R_p} = \frac{1}{R_1} + \frac{1}{R_2}$$
$$\frac{1}{R_p} = \frac{1}{40\,\Omega} + \frac{1}{60\,\Omega}$$

This means R_p is 24 Ω. The total resistance of the circuit is thus

$$R_s = R_1 + R_2 = 24\,\Omega + 20\,\Omega \text{ or } 44\,\Omega.$$

(b) Because $P = IV$ and $V = IR$ we can substitute in to obtain

$$P = IV$$
$$P = \frac{V}{R}V$$
$$P = \frac{V^2}{R}$$
$$P = \frac{(12\text{ V})^2}{44\,\Omega}$$

or $P = 3.27$ W

(c) The energy transferred in 15 minutes is given by:

$$E = Pt$$
$$E = (3.3\text{ W}) \times 15\min\left(\frac{60\text{ s}}{\min}\right)$$
$$= 2970\text{ J}$$

(d) The battery will drain more quickly and discharge in less than 15 minutes. If the resistors are all in parallel, the total equivalent resistance will be lower, so I will be higher and $P = IV$ shows that power will be higher. With a higher power, the charges will drain from the battery more quickly.

4. (a) We'll consider two points in the flow stream: Point 1 at the surface of the water in the tank and Point 2 at the position of the hole. Because the cross-sectional area of the tank is so large, we can apply Torricelli's Theorem,

$$v = \sqrt{2g(y_1 - y_2)}$$

where y_1 is the vertical distance from the ground to Point 1 and y_2 is the vertical distance from the ground to Point 2. We conclude that the efflux speed of the water is

$$v = \sqrt{2(10 \text{ m/s}^2)(50\text{m} - 35\text{m})} = \sqrt{2(10 \text{ m/s}^2)(15\text{m})} = 17.3 \text{ m/s}$$

(b) The flow rate is equal to the cross-sectional area times the flow speed: $f = Av$. Using the result of part (a), we first find that

$$A = \frac{f}{v} = \frac{3 \times 10^{-4} \frac{\text{m}^3}{\text{min}} \cdot \frac{1 \text{ min}}{60 \text{ s}}}{17.3 \text{ m/s}} = 2.89 \times 10^{-7} \text{ m}^2$$

Since the area of the hole is given by $A = \pi r^2$, we find that the radius of the hole is

$$r = \sqrt{\frac{A}{\pi}} = \sqrt{\frac{2.89 \times 10^{-7} \text{ m}^2}{\pi}} = 3.03 \times 10^{-4} \text{ m}$$

Therefore, the diameter of the hole is twice this value: $d = 2r = 6.06 \times 10^{-4}$ m.

(c) The horizontal distance, x, will be equal to vt, where v is the efflux speed we found in part (a) and t is the time it takes for water to reach the ground. Because the stream of water has zero vertical speed when it exits the hole, we can use the equation $y = \frac{1}{2}at^2$, with $a = g$, to find t. This gives

$$t = \sqrt{\frac{2y}{g}}$$

Since the vertical distance the water drops after exiting the hole is $y = 35$ m, the horizontal distance we're looking for is

$$x = v\sqrt{\frac{2y}{g}} = (17.3 \text{ m/s})\sqrt{\frac{2(35 \text{ m})}{10 \text{ m/s}^2}} = 45.8 \text{ m}$$

Part III
About the
AP Physics 2
Exam

- The Structure of the AP Physics 2 Exam
- How the AP Exams Are Used
- Other Resources
- Designing Your Study Plan

THE STRUCTURE OF THE AP PHYSICS 2 EXAM

The AP Physics 2 Exam consists of two sections: a multiple-choice section and a free-response section. Questions in the multiple-choice section are each followed by four possible responses. In the single-select subsection, only one of these possible answers is correct. In the multi-select subsection, two of these possible answers are correct. It is your job to choose the right answer. Each correct answer is worth one point.

Section	Timing	Scoring	Question Type	Number of Questions
I: Multiple Choice	90 minutes	50% of exam score	Single-select (discrete questions and questions in sets with one correct answer)	45
			Multi-select (discrete questions with two correct answers)	5
				Total—50

Section	Timing	Scoring	Question Type	Number of Questions
II: Free Response	90 minutes	50% of exam score	Experimental Design	1
			Qualitative/ Quantitative/ Translation	1
			Short Answer	2
				Total—4

The free-response section requires you to write out your solutions, showing your work. The total amount of time for this section is 90 minutes, so you have an average of 22.5 minutes per question. Unlike the multiple choice section, which is scored by computer, the free-response section is graded by high school and college teachers. They have guidelines for awarding partial credit, so you don't need to correctly answer every part to get points. You are allowed to use a calculator on the entire AP Physics 2 Exam (programmable or graphing calculators are okay, but ones with a typewriter-style keyboard are not) and a table of equations is provided for your use. The two sections—multiple choice and free response—are weighted equally, so each is worth 50 percent of your grade.

Grades on the AP Physics 2 Exam are reported as a number: either 1, 2, 3, 4, or 5. The descriptions for each of these five numerical scores are as follows:

AP Exam Grade	Description
5	Extremely well qualified
4	Well qualified
3	Qualified
2	Possibly qualified
1	No recommendation

Colleges are generally looking for a 4 or 5, but some may grant credit for a 3. How well do you have to do to earn such a grade? Each test is curved, and specific cut-offs for each grade vary a little from year to year, but here's a rough idea of how many points you must earn—as a percentage of the maximum possible raw score—to achieve each of the grades 2 through 5:

AP Exam Grade	Percentage Needed
5	≥ 75%
4	≥ 60%
3	≥ 45%
2	≥ 35%

AP Physics 2: Algebra-Based Course Content

You may be using this book as a supplementary text as you take an AP Physics 2 course at your high school or you may be using it on your own. The College Board is very detailed in what they require your AP teacher to cover in his or her AP Physics 2 course. They explain the course that you should know with the following summary.

Students explore principles of fluids, thermodynamics, electricity, magnetism, optics, and topics in modern physics. The course is based on seven Big Ideas, which encompass core scientific principles, theories, and processes that cut across traditional boundaries and provide a broad way of thinking about the physical world.

The following are Big Ideas:

- Objects and systems have properties such as mass and charge. Systems may have internal structure.
- Fields existing in space can be used to explain interactions.
- The interactions of an object with other objects can be described by forces.
- Interactions between systems can result in changes in those systems.
- Changes that occur as a result of interactions are constrained by conservation laws.

- Waves can transfer energy and momentum from one location to another without the permanent transfer of mass and serve as a mathematical model for the description of other phenomena.
- The mathematics of probability can be used to describe the behavior of complex systems and to interpret the behavior of quantum mechanical systems.

Naturally, it's important to be familiar with the topics—to understand the basics of the theory, to know the definitions of the fundamental quantities, and to recognize and be able to use the equations. Then, you must acquire practice at applying what you've learned to answering questions like you'll see on the exam. This book is designed to review all of the content areas covered on the exam, illustrated by hundreds of examples. Also, each content review chapter is followed by practice multiple-choice and free-response questions, and perhaps even more important, *answers and explanations are provided for every example and question in this book.* You'll learn as much—if not more—from actively reading the solutions as you will from reading the text and examples. Also, two full-length practice tests (with solutions) are provided in this book. The difficulty level of the examples and questions in this book is at or slightly above AP level, so if you have the time and motivation to attack these questions and learn from the solutions, you should feel confident that you can do your very best on the real thing.

The College Board also gives us a list of Physics 2 Principles that they describe as foundational physics principles. If you are taking an AP Physics 2 course, you will likely recognize these categories from your textbook sections. Be sure that you're familiar with the following principles as you prepare for your exam:

- Thermodynamics: Laws of Thermodynamics, Ideal Gases, Kinetic Theory
- Fluid Statics and Dynamics
- Electrostatics: Electric Force, Electric Field, and Electric Potential
- DC Circuits and RC Circuits (steady-state only)
- Magnetism and Electromagnetic Induction
- Geometric and Physical Optics
- Quantum Physics, Atomic, and Nuclear Physics

HOW AP EXAMS ARE USED

Different colleges use AP Exams in different ways, so it is important that you go to a particular college's web site to determine how it uses AP Exams. The three items below represent the main ways in which AP Exam scores can be used:

- **College Credit.** Some colleges will give you college credit if you score well on an AP Exam. These credits count toward your graduation requirements, meaning that you can take fewer courses while in college. Given the cost of college, this could be quite a benefit, indeed.
- **Satisfy Requirements.** Some colleges will allow you to "place out" of certain requirements if you do well on an AP Exam, even if they do not give you actual college credits. For example, you might not need to take an introductory-level course, or perhaps you might not need to take a class in a certain discipline at all.
- **Admissions Plus.** Even if your AP Exam will not result in college credit or even allow you to place out of certain courses, most colleges will respect your decision to push yourself by taking an AP Course or even an AP Exam outside of a course. A high score on an AP Exam shows mastery of more difficult content than is taught in many high school courses, and colleges may take that into account during the admissions process.

Are You Preparing for College?
Check out all of the useful books from The Princeton Review, including *The Best 381 Colleges, Cracking the SAT, Cracking the ACT,* and more!

OTHER RESOURCES

There are many resources available to help you improve your score on the AP Physics 2 Exam, not the least of which are your **teachers**. If you are taking an AP class, you may be able to get extra attention from your teacher, such as obtaining feedback on your essays. If you are not in an AP course, reach out to a teacher who teaches AP Physics 2 and ask if the teacher will review your essays or otherwise help you with content.

Another wonderful resource is **AP Students**, the official site of the AP Exams. The scope of the information at this site is quite broad and includes:

- a Course Description, which includes details on what content is covered and sample questions
- sample questions from the AP Physics 2 Exam

The AP Students home page address is: **http://apstudent.collegeboard.org**.

Finally, The Princeton Review offers tutoring for the AP Physics 2 Exam. Our expert instructors can help you refine your strategic approach and add to your content knowledge. For more information, call 1-800-2REVIEW.

Study Breaks Are Important
Don't burn yourself out before test day. Remember to take breaks every so often—go for a walk, listen to a favorite album, get some fresh air.

DESIGNING YOUR STUDY PLAN

As part of the Introduction, you identified some areas of potential improvement. Let's now delve further into your performance on Test 1, with the goal of developing a study plan appropriate to your needs and time commitment.

Read the answers and explanations associated with the Multiple Choice questions (starting at page 29). After you have done so, respond to the following questions:

- Review the list of Big Ideas on pages 43–44. Next to each topic, indicate your rank of the topic as follows: "1" means "I need a lot of work on this," "2" means "I need to beef up my knowledge," and "3" means "I know this topic well."

- How many days/weeks/months away is your exam?

- What time of day is your best, most focused study time?

- How much time per day/week/month will you devote to preparing for your exam?

- When will you do this preparation? (Be as specific as possible: Mondays and Wednesdays from 3:00 P.M. to 4:00 P.M., for example)

- Based on the answers above, will you focus on strategy (Part Two) or content (Part Three) or both?

- What are your overall goals in using this book?

Part IV
Test-Taking
Strategies for the
AP Physics 2
Exam

PREVIEW

Review your Practice Test 1 results and then respond to the following questions:

- How many multiple-choice questions did you miss even though you knew the answer?

- On how many multiple-choice questions did you guess blindly?

- How many multiple-choice questions did you miss after eliminating some answers and guessing based on the remaining answers?

- Did you find any of the free-response questions easier or harder than the others—and, if so, why?

HOW TO USE THE CHAPTERS IN THIS PART

For the following Strategy chapters, think about what you are doing now before you read the chapters. As you read and engage in the directed practice, be sure to appreciate the ways you can change your approach.

Chapter 1
How to Approach
Multiple-Choice
Questions

CRACKING THE MULTIPLE-CHOICE SECTION

All the multiple-choice questions will have a similar format: Each will be followed by four answer choices. At times, it may seem that there could be more than one possible correct answer. For the single choice problems, which represent 45 out of your 50 multiple choice problems, there is only one correct answer! For the last 5 problems of the multiple choice section, there will be two correct answers per question. Keep in mind that this section is identified before you start it. Answers resulting from common mistakes are often included in the four answer choices to trap you.

Use the Answer Sheet

For the multiple-choice section, you write the answers not in the test booklet but on a separate answer sheet (very similar to the ones we've supplied at the very end of this book). Four oval-shaped answer bubbles follow the question number, one for each possible answer. Don't forget to fill in all your answers on the answer sheet. Don't just mark them in the test booklet. Marks in the test booklet will not be graded. Also, make sure that your filled-in answers correspond to the correct question numbers! Check your answer sheet after every five answers to make sure you haven't skipped any bubbles by mistake.

The Two-Pass System

The AP Physics 2 Exam covers a broad range of topics. There's no way, even with our extensive review, that you will know everything about every topic in algebra-based physics. So, what should you do?

Adopt a two-pass system. The two-pass system entails going through the test and answering the easy questions first. Save the more time-consuming questions for later. (Don't worry—you'll have time to do them later!) First, read the question and decide if it is a "now" or "later" question. If you decide this is a "now" question, answer it in the test booklet. If it is a "later" question, come back to it. Once you have finished all the "now" questions on a double page, transfer the answers to your bubble sheet. Flip the page and repeat the process.

Once you've finished all the "now" questions, move on to the "later" questions. Start with the easier questions first. These are the ones that require calculations or that require you to eliminate the answer choices (in essence, the correct answer does not jump out at you immediately). Transfer your answers to your bubble sheet as soon as you answer these "later" questions.

Watch Out for Those Bubbles!

Because you're skipping problems, you need to keep careful track of the bubbles on your answer sheet. One way to accomplish this is by answering all the questions on a page and then transferring your choices to the answer sheet. If you prefer to enter them one by one, make sure you double-check the number beside the ovals before filling them in. We'd hate to see you lose points because you forgot to skip a bubble!

Process of Elimination (POE)

On most tests, you need to know your material backward and forward in order to get the right answer. In other words, if you don't know the answer beforehand, you probably won't answer the question correctly. This is particularly true of fill-in-the-blank and essay questions. We're taught to think that the only way to get a question right is by knowing the answer. However, that's not the case on Section I of the AP Physics 2 Exam. You can get a perfect score on this portion of the test without knowing a single right answer—provided you know all the wrong answers!

What are we talking about? This is perhaps the most important technique to use on the multiple-choice section of the exam. Let's take a look at an example.

41. A gas undergoes an expansion-compression cycle. If, plotted on a P–V diagram, the cycle is counterclockwise and the work is 300 J in magnitude, what was the heat transfer during this cycle?

    (A) 300 J into the system
    (B) 300 J out of the system
    (C) 600 J into the system
    (D) 600 J out of the system

Now, if this were a fill-in-the-blank-style question, you might be in a heap of trouble. But let's take a look at what we've got. If we can eliminate a few of the answer choices, we can get closer to making an educated guess.

For a cycle, $\Delta U = 0$, so we know the magnitude of the work and heat transfer have to be the same. This eliminates (C) and (D). Since the cycle is counterclockwise on a P–V diagram, the work has to be positive, so the heat transfer is negative. A negative heat transfer is heat leaving the system. Choice (B) is correct.

We think we've illustrated our point: Process of Elimination is the best way to approach the multiple-choice questions. Even when you don't know the answer right off the bat, you'll surely know that two or three of the answer choices are not correct. What then?

Aggressive Guessing

As mentioned earlier, you are scored only on the number of questions you get right, so we know guessing can't hurt you. But can it help you? It sure can. Let's say you guess on five questions; odds are you'll get one right. So you've already increased your score by one point. Now, let's add POE into the equation. If you can eliminate as many as two answer choices from each question, your chances of getting them right increase, and so does your overall score. Remember: Don't leave any bubbles blank on test day!

General Advice

Answering 50 multiple-choice questions in 90 minutes can be challenging. Make sure to pace yourself accordingly and remember that you do not need to answer every question correctly to do well. Exploit the multiple-choice structure of this section. For 45 of your 50 multiple-choice questions, there are three wrong answers and only a single correct one (for the multi-correct section, the last 5 questions, there are two wrong and two correct), so even if you don't know exactly which one is the right answer, you can eliminate some that you know for sure are wrong. Then you can make an educated guess from among the answers that are left and greatly increase your odds of getting that question correct.

Problems with graphs and diagrams are usually the fastest to solve and problems with an explanation for each answer usually take the longest to work through. Do not spend too much time on any one problem or you may not get to easier problems further into the test.

These practice exams are written to give you an idea of the format of the test, the difficulty of the questions, and to allow you to practice pacing yourself. Take them in the same circumstances as you will encounter during the real exam.

Reflect

Respond to the following questions:

- How long will you spend on multiple choice questions?

- How will you change your approach to multiple choice questions?

- What is your multiple choice guessing strategy?

- Will you seek further help, outside of this book (such as a teacher, tutor, or AP Students), on how to approach the questions that you will see on the AP Physics 2 Exam?

Chapter 2
How to Approach
Free-Response
Questions

FREE-RESPONSE SECTION

On the free-response section, be sure to show the graders what you're thinking. Write clearly—that is *very* important—and show your steps. If you make a mistake in one part and carry an incorrect result to a later part of the question, you can still earn valuable points if your method is correct. But the graders cannot give you credit for work they can't follow or can't read. And, where appropriate, be sure to include units on your final answers.

The most important advice we can give you for the free-response section of the AP Physics 2 Exam is to read the questions carefully and answer according to exactly what the questions are asking you to do. Credit for the answers depends not only on the quality of the solutions but also on how they are explained. On the AP Physics 2 Exam, the words "justify," "explain," "calculate," "what is," "determine," and "derive" have specific meanings, and the graders are looking for very precise approaches in your explanations in order to get maximum credit.

Questions that ask you to "justify" are looking for you both to show an understanding in words of the principles underlying physical phenomena and to perform the mathematical operations needed to arrive at the correct answer. The word "justify" as well as the word "explain" requires that you support your answers with text, equations, calculations, diagrams, or graphs. In some cases, the text or equations must elucidate physics fundamentals or laws, while in other cases they will serve to analyze the behavior of different values or different types of variables in the equation.

The word "calculate" requires you to show numerical or algebraic work to arrive at the final answer. In contrast, "what is" and "determine" questions signify that full credit may be given without showing mathematical work. Just remember, showing work that leads to the correct answer is always a good idea when possible, especially since showing work may still earn you partial credit even if the answer is not correct.

"Derive" questions are looking for a more specific approach, which entails beginning the solution with one or more fundamental equations and then arriving at the final answer through the proper use of mathematics, usually involving some algebra.

CRACKING THE FREE-RESPONSE SECTION

Section II is worth 50 percent of your grade on the AP Physics 2 Exam. This section is composed of four free-response questions. You're given a total of 90 minutes for this section. We at The Princeton Review recommend that you spend 50 minutes on the first two questions (experimental design/translation) and then 40 minutes on the second two questions (the short response questions) because short response questions should require a bit less time. Pacing is a personal decision, though, so feel free to build your own timeline for the free-response section.

Clearly Explain and Justify Your Answers

Remember that your answers to the free-response questions are graded by readers and not by computers. Communication is a very important part of AP Physics 2. Compose your answers in precise sentences. Just getting the correct numerical answer is not enough. You should be able to explain your reasoning behind the technique that you selected and communicate your answer in the context of the problem. Even if the question does not explicitly say so, always explain and justify every step of your answer, including the final answer. Do not expect the graders to read between the lines. Explain everything as though somebody with no knowledge of physics is going to read it. Be sure to present your solution in a systematic manner using solid logic and appropriate language. And remember: Although you won't earn points for neatness, the graders can't give you a grade if they can't read and understand your solution!

Use Only the Space You Need

Do not try to fill up the space provided for each question. The space given is usually more than enough. The people who design the tests realize that some students write in big letters and some students make mistakes and need extra space for corrections. So if you have a complete solution, don't worry about the extra space. Writing more will not earn you extra credit. In fact, many students tend to go overboard and shoot themselves in the foot by making a mistake after they've already written the right answer.

Read the Whole Question!

Some questions might have several subparts. Try to answer them all, and don't give up on the question if one part is giving you trouble. For example, if the answer to part (b) depends on the answer to part (a), but you think you got the answer to part (a) wrong, you should still go ahead and do part (b) using your answer to part (a) as required. Chances are that the grader will not mark you wrong twice, unless it is obvious from your answer that you should have discovered your mistake.

Use Common Sense

Always use your common sense in answering questions. For example, on one free-response question that asked students to compute the mean weight of newborn babies from given data, some students answered 70 pounds. It should have been immediately obvious that the answer was probably off by a decimal point. A 70-pound baby would be a giant! This is an important mistake that should be easy to fix. Some mistakes may not be so obvious from the answer. However, the grader will consider simple, easily recognizable errors to be very important.

Think Like a Grader

When answering questions, try to think about what kind of answer the grader is expecting. Look at past free-response questions and grading rubrics on the College Board website. These examples will give you some idea of how the answers should be phrased. The graders are told to keep in mind that there are two aspects to the scoring of free-response answers: showing statistical knowledge and communicating that knowledge. Again, responses should be written as clearly as possible in complete sentences. You don't need to show all the steps of a calculation, but you must explain how you got your answer and why you chose the technique you used.

Think Before You Write

Abraham Lincoln once said that if he had eight hours to chop down a tree, he would spend six of them sharpening his axe. Be sure to spend some time thinking about what the question is, what answers are being asked for, what answers might make sense, and what your intuition is before starting to write. These questions aren't meant to trick you, so all the information you need is given. If you think you don't have the right information, you may have misunderstood the question. In some calculations, it is easy to get confused, so think about whether your answers make sense in terms of what the question is asking. If you have some idea of what the answer should look like before starting to write, then you will avoid getting sidetracked and wasting time on dead-ends.

Reflect

Respond to the following questions:

- How much time will you spend on the short free-response questions? What about the long free-response questions?

- What will you do before you begin writing your free response answers?

- Will you seek further help, outside of this book (such as a teacher, tutor, or AP Students), on how to approach the questions that you will see on the AP Physics 2 Exam?

Part V
Content Review
for the AP
Physics 2 Exam

Chapter 3
Fluid Mechanics

INTRODUCTION

In this chapter, we'll discuss some of the fundamental concepts dealing with substances that can flow, which are known as **fluids**. The term *fluid* refers to both liquids and gases. While there are distinctions between liquids and gases, this chapter focuses on the similarities of all fluids.

DENSITY

Although the concept of *mass* is central to your study of mechanics (because of the all-important equation $F_{net} = ma$), it is the substance's *density* that turns out to be more useful in fluid mechanics.

By definition, the density of a substance is its mass per unit volume, and it's typically denoted by the letter ρ (the Greek letter *rho*):

Equation Sheet

$$\text{density} = \frac{\text{mass}}{\text{volume}}$$
$$\rho = \frac{m}{V}$$

Note that this equation immediately implies that m is equal to ρV when ρ is a constant.

For example, if 10^{-3} m³ of oil has a mass of 0.8 kg, then the density of this oil is

$$\rho = \frac{m}{V} = \frac{0.8 \text{ kg}}{10^{-3} \text{ m}^3} = 800 \frac{\text{kg}}{\text{m}^3}$$

PRESSURE

If we place an object in a fluid, the fluid exerts a contact force on the object. How that force is distributed over any small area of the object's surface defines the **pressure**:

Equation Sheet

$$\text{pressure} = \frac{\text{force}_\perp}{\text{area}}$$
$$P = \frac{F_\perp}{A}$$

The subscript ⊥ (which means *perpendicular*) is meant to emphasize that the pressure is defined to be the magnitude of the force that acts perpendicular to the surface, divided by the area. Because force is measured in newtons (N) and area is expressed in square meters (m²), the SI unit for pressure is the newton per square meter. This unit is given its own name: the **pascal**, abbreviated Pa:

$$\text{SI unit of pressure: } 1 \text{ pascal} = 1 \text{ Pa} = 1\frac{\text{N}}{\text{m}^2}$$

One pascal is a very tiny amount of pressure; for example, a nickel on a table exerts about 140 Pa of pressure, just due to its weight alone. For this reason, you'll often see pressures expressed in kPa (kilopascals, where 1 kPa = 10^3 Pa) or even in MPa (megapascals, where 1 MPa = 10^6 Pa). Another common unit for pressure is the atmosphere (atm). At sea level, atmospheric pressure, P_{atm}, is about 101,300 Pa; this is 1 **atmosphere**.

Example 1 A vertical column made of cement has a base area of 0.5 m². If its height is 2 m, and the density of cement is 3000 kg/m³, how much pressure does this column exert on the ground?

Solution. The force the column exerts on the ground is equal to its weight, *mg*, so we'll find the pressure it exerts by dividing this by the base area, *A*. The mass of the column is equal to ρV, which we calculate as follows:

$$m = \rho V = \rho A h = (3 \times 10^3 \, \frac{\text{kg}}{\text{m}^3})(0.5 \text{ m}^2)(2 \text{ m}) = 3 \times 10^3 \text{ kg}$$

Therefore,

$$P = \frac{F}{A} = \frac{mg}{A} = \frac{(3 \times 10^3 \text{ kg})(10 \, \frac{\text{N}}{\text{kg}})}{0.5 \text{ m}^2} = 6 \times 10^4 \text{ Pa} = 60 \text{ kPa}$$

Now, you may be used to the unit m/s² for g. However, for this chapter, it is simpler to use the unit N/kg for g, as it is equivalent to m/s². So don't be confused when you see it later on.

HYDROSTATIC PRESSURE

Imagine that we have a tank with a lid on top, filled with some liquid. Suspended from this lid is a string, attached to a thin sheet of metal which hangs horizontally. The figures below show two views of this:

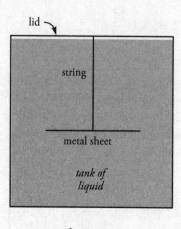

front view

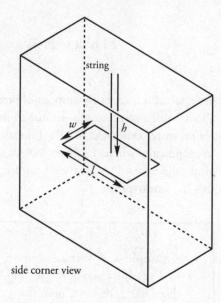

side corner view

The weight of the liquid produces a force that pushes down on the metal sheet. If the sheet has length l and width w, and is at depth h below the surface of the liquid, then the weight of the liquid on top of the sheet is

$$F_g = mg = \rho V g = \rho(lwh)g$$

where ρ is the liquid's density. If we divide this weight by the area of the sheet ($A = lw$), we get the pressure due to the liquid:

$$P_{\text{liquid}} = \frac{\text{force}}{\text{area}} = \frac{F_{g\,\text{liquid}}}{A} = \frac{\rho(lwh)g}{lw} = \rho gh$$

Since the liquid is at rest, this is known as **hydrostatic pressure**.

Note that the hydrostatic pressure due to the liquid, $P_{\text{liquid}} = \rho gh$, depends only on the density of the liquid and the depth below the surface; in fact, it's proportional to both of these quantities. One important consequence of this is that the shape of the container doesn't matter. For example, if all the containers in the figure below are filled with the same liquid, then the pressure is the same at every point along the horizontal dashed line (and within a container), simply because every point on this line is at the same depth, h, below the surface of the liquid.

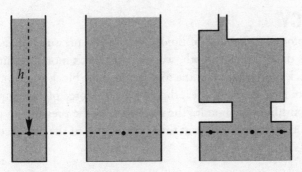

If the liquid in the tank were open to the atmosphere, then the total (or absolute) pressure at depth h would be equal to the pressure pushing down on the surface—the atmospheric pressure, P_{atm}—plus the pressure due to the liquid alone. In general, the pressure on the surface of the liquid is some pressure P_0 and the total pressure is

$$\text{total (absolute) pressure: } P_{total} = P_0 + P_{liquid} = P_0 + \rho g h$$

When the pressure at the top of the liquid is atmospheric pressure (as in the scenario above), then our equation for total pressure reduces to

$$P_{total} = P_{atm} + \rho g h$$

Because pressure is the *magnitude* of the force per area, pressure is a scalar. It has no direction. The direction of the force due to the pressure on any small surface is perpendicular to that surface. For example, in the figure below, the pressure at Point A is the same as the pressure at Point B, because they're at the same depth.

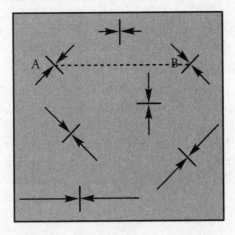

But, as you can see, the direction of the force due to the pressure varies depending on the orientation of the surface—and even which side of the surface—the force is pushing on.

It's All About Depth
Pressure is proportional to depth of an object, not the mass of that object. An elephant and a feather at the same depth feel the same amount of preasure. How they react to this pressure is a different story.

BUOYANCY

Let's place a block in our tank of fluid. Because the pressure on each side of the block depends on its average depth, we see that there's more pressure on the bottom of the block than there is on the top. Because the block is rectangular and the top and bottom have the same area, there's a greater force pushing up on the block than there is pushing down on it. The forces due to the pressure on the other four sides cancel out (because they are at the same depth), so the net force on the block is upward.

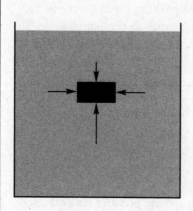

front view

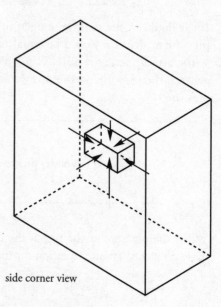

side corner view

This net upward force is called the **buoyant force** (or just **buoyancy** for short), denoted $\mathbf{F}_{buoy}$. We calculate the magnitude of the buoyant force using **Archimedes' principle**; in words, Archimedes' principle says

> *The strength of the buoyant force on the object is equal to the weight of the fluid displaced by the object.*

When an object is partially or completely submerged in a fluid, the volume of the object submerged, which we call V_{sub}, is the volume of the fluid displaced. By multiplying this volume by the density of the fluid, we get the mass of the fluid displaced; then, multiplying this mass by g gives us the weight of the fluid displaced. So, here's Archimedes' principle as a mathematical equation:

> Buoyant force: $F_{buoy} = \rho_{fluid} V_{sub} g$

When an object floats, its submerged volume is just enough to make the buoyant force it feels balance its weight. So, if an object's density is ρ_{object} and its (total) volume is V, its weight will be $mg = \rho V g$. The buoyant force it feels is $\rho_{\text{fluid}} V_{\text{sub}} g$. Setting these equal to each other, we find that

$$\frac{V_{\text{sub}}}{V} = \frac{\rho_{\text{object}}}{\rho_{\text{fluid}}}$$

So, if $\rho_{\text{object}} < \rho_{\text{fluid}}$, then the object will float; and the fraction of its volume submerged is the same as the ratio of its density to the fluid's density. For example, if the object's density is 2/3 the density of the fluid, then the object will float, and 2/3 of the object will be submerged.

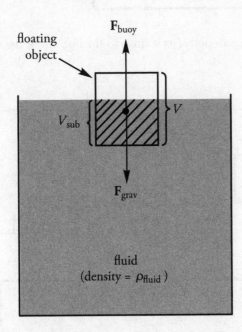

If an object is denser than the fluid, it will sink. In this case, even if the entire object is submerged (in an attempt to maximize V_{sub} and the buoyant force), its weight is still greater than the buoyant force, and down it goes. And if an object just happens to have the same density as the fluid, it will be happy hovering (in static equilibrium) anywhere underneath the fluid.

Let's summarize, and go through the steps of an object floating and then an object as it sinks and hits the bottom of a tank. In the diagram on the next page we have an object that is floating in a liquid.

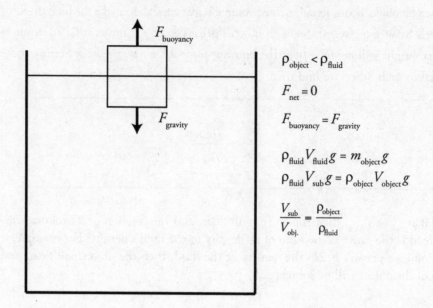

$$\rho_{object} < \rho_{fluid}$$

$$F_{net} = 0$$

$$F_{buoyancy} = F_{gravity}$$

$$\rho_{fluid} V_{fluid} g = m_{object} g$$
$$\rho_{fluid} V_{sub} g = \rho_{object} V_{object} g$$

$$\frac{V_{sub}}{V_{obj.}} = \frac{\rho_{object}}{\rho_{fluid}}$$

Below is a diagram of an object as it sinks to the bottom of the tank.

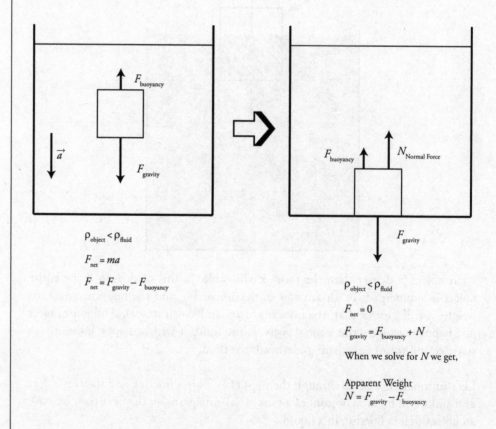

$$\rho_{object} < \rho_{fluid}$$

$$F_{net} = ma$$

$$F_{net} = F_{gravity} - F_{buoyancy}$$

$$\rho_{object} < \rho_{fluid}$$

$$F_{net} = 0$$

$$F_{gravity} = F_{buoyancy} + N$$

When we solve for N we get,

Apparent Weight
$$N = F_{gravity} - F_{buoyancy}$$

Because the buoyant force is always directed upwards, and the force of gravity is defined to point downwards, these forces are always directed in opposite directions. The result of the buoyant force opposing gravity is that any object in a fluid appears to weigh less than when it is not in the fluid, and as the fluid density increases, the apparent weight decreases. The apparent weight of an object is commonly referred to as the normal force of that object.

Example 2 An object with a mass of 150 kg and a volume of 0.75 m³ is floating in ethyl alcohol, whose density is 800 kg/m³. What fraction of the object's volume is above the surface of the fluid?

Solution. The density of the object is

$$\rho_{object} = \frac{m}{V} = \frac{150 \text{ kg}}{0.75 \text{ m}^3} = 200 \, \frac{\text{kg}}{\text{m}^3}$$

The ratio of the object's density to the fluid's density is

$$\frac{\rho_{object}}{\rho_{fluid}} = \frac{200 \text{ kg/m}^3}{800 \text{ kg/m}^3} = \frac{1}{4}$$

This means that 1/4 of the object's volume is *below* the surface of the fluid; therefore, the fraction *above* the surface is 1 – (1/4) = 3/4. Make sure you know exactly what the question is asking. Typically you use this equation to find the volume submerged, but this question (Example 2) asks for the volume not submerged.

Example 3 A brick, of density 2000 kg/m³ and volume 1.5 × 10⁻³ m³, is dropped into a swimming pool full of water.
        (a) Explain briefly why the brick will sink.
        (b) When the brick is lying on the bottom of the pool, what is the apparent weight of the brick?

Solution.

(a) The brick has a greater density than the surrounding fluid (water), so it will sink.

(b) When the brick is lying on the bottom surface of the pool, it is totally submerged, so $V_{sub} = V$; this means the buoyant force on the brick is

$$F_{buoy} = \rho_{fluid}V_{sub}g = \rho_{water}Vg$$
$$= (1000 \text{ kg/m}^3)(1.5 \times 10^{-3} \text{ m}^3)(10 \text{ N/kg})$$
$$= 15 \text{ N}$$

The weight of the brick is

$$F_g = mg = \rho_{brick} V g$$
$$= (2000 \text{ kg/m}^3)(1.5 \times 10^{-3} \text{ m}^3)(10 \text{ N/kg})$$
$$= 30 \text{ N}$$

When the brick is lying on the bottom of the pool, the net force it feels is zero.

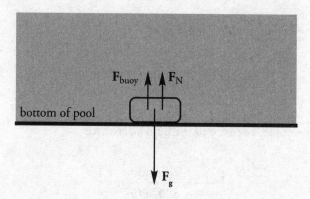

Therefore, we must have $F_{buoy} + F_N = F_g$, so $F_N = F_g - F_{buoy} = 30 \text{ N} - 15 \text{ N} = 15 \text{ N}$. So the apparent weight of the brick, that is, the normal force it feels, is 15 N.

Example 4 A helium balloon has a volume of 0.03 m³. What is the net force on the balloon if it's surrounded by air? (Note: The density of helium is 0.2 kg/m³, and the density of air is 1.2 kg/m³.)

Solution. The balloon will feel a buoyant force upward, and the force of gravity—the balloon's weight—downward. Because the balloon is completely surrounded by air, $V_{sub} = V$, and the buoyant force is

$$F_{buoy} = \rho_{fluid} V_{sub} g = \rho_{air} V g$$
$$= (1.2 \text{ kg/m}^3)(0.03 \text{ m}^3)(10 \text{ N/kg})$$
$$= 0.36 \text{ N}$$

The weight of the balloon is

$$F_g = mg = \rho_{helium} V g = (0.2 \text{ kg/m}^3)(0.03 \text{ m}^3)(10 \text{ N/kg})$$
$$= 0.06 \text{ N}$$

Equation Sheet

Because $F_{buoy} > w$, the net force on the balloon is upward and has magnitude

$$F_{net} = F_{buoy} - F_g = 0.36 \text{ N} - 0.06 \text{ N} = 0.3 \text{ N}$$

This is why a helium balloon floats away if you let go of its string.

FLOW RATE AND THE CONTINUITY EQUATION

Consider a pipe through which fluid is flowing. The **flow rate**, *f*, is the volume of fluid that passes a particular point per unit time. In SI units, flow rate is expressed in m³/s. To find the flow rate, all we need to do is multiply the size of the pipe at a particular location by the average speed of the flow at that point. In this context, the size of the pipe means the area of a cross-section of the pipe that is perpendicular to the direction of the flow.

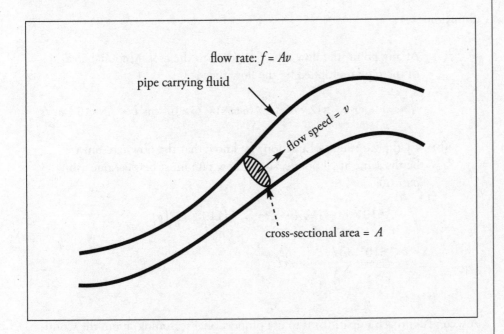

flow rate: $f = Av$

pipe carrying fluid

flow speed = v

cross-sectional area = A

Be careful not to confuse *flow rate* with *flow speed*; flow rate tells us how *much* fluid flows per unit time; flow speed tells us how *fast* it's moving. The flow rate is measured in cubic meters per second, while the flow speed is measured in meters per second.

If the pipe is carrying a fluid that is incompressible—that is, its density cannot be changed—then the flow rate must be the same everywhere along the pipe. A liquid is typically an incompressible fluid, whereas a gas is compressible. Rewriting this using $f = Av$, we get the **Continuity Equation:**

$$A_1 v_1 = A_2 v_2$$

Equation Sheet

Because the product Av is a constant, the flow speed will increase where the pipe narrows, (decrease in area) and decrease where the pipe widens (increase in area). In fact, we can say that the flow speed is inversely proportional to the cross-sectional area—or to the square of the radius (in the case of a circular cross-section)—of the pipe.

> **Example 5** A circular pipe of non-uniform diameter carries water. At one point in the pipe, the radius is 2 cm and the flow speed is 6 m/s.
>
> (a) What is the flow rate?
>
> (b) What is the flow speed at a point where the pipe constricts to a radius of 1 cm?

Solution.

(a) At any point, the flow rate, f, is equal to the cross-sectional area of the pipe multiplied by the flow speed

$$f = Av = \pi r^2 v = \pi (2 \times 10^{-2} \text{ m})^2 (6 \text{ m/s}) \approx 75 \times 10^{-4} \text{ m}^3/s = 7.5 \times 10^{-3} \text{ m}^3/s$$

(b) By the Continuity Equation, we know that the flow rate must be the same at all points, so the flow rate must be the same as in part (a).

$$f = 7.510^{-3} \text{ m}^3/s = Av = \pi r^2 = \pi \left(110^{-2} m\right)^2 (v)$$

$$v = \frac{7.510^{-3} m^3/s}{\pi \left(110^{-2} m\right)^2} \approx 24 m/s$$

A second approach to part (b) is to use proportional reasoning. From the Continuity Equation, we know that v, the flow speed, is inversely proportional to A, the cross-sectional area of the pipe. If the pipe's radius decreases by a factor of 2 (from 2 cm to 1 cm), then A decreases by a factor of 4, because A is proportional to r^2.

If A decreases by a factor of 4, then v will increase by a factor of 4. Therefore, the flow speed at a point where the pipe's radius is 1 cm will be $4 \cdot (6 \ m/s) = 24 \ m/s$.

BERNOULLI'S EQUATION

The most important equation in fluid mechanics is **Bernoulli's Equation**, which is the statement of conservation of energy for ideal fluid flow. First, let's describe the conditions that make fluid flow *ideal*.

- *The fluid is incompressible.*

  This works very well for liquids and also applies to gases if the pressure changes are small.

- *The fluid's viscosity is negligible.*

  Viscosity is the force of cohesion between molecules in a fluid; think of viscosity as internal friction for fluids. For example, maple syrup is sticky and has a greater viscosity than water: there's more resistance to a flow of maple syrup than to a flow of water. While Bernoulli's Equation would give good results when applied to a flow of water, it would not give good results if it were applied to a flow of maple syrup.

- *The flow is streamline.*

  In a tube carrying a flowing fluid, a **streamline** is just what it sounds like: it's a *line* in the *stream*. If we were to inject a drop of dye into a clear glass pipe carrying, say, water, we'd see a streak of dye in the pipe, indicating a streamline.

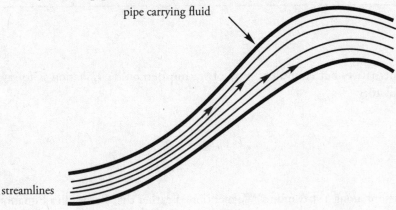

pipe carrying fluid

streamlines

When the flow is streamline, the fluid moves smoothly through the tube. (The opposite of streamline flow is **turbulent** flow, which is characterized by rapidly swirling whirlpools; such chaotic flow is unpredictable.)

If the three conditions described above hold, and the flow rate, f, is steady (meaning it doesn't change with time), Bernoulli's Equation can be applied to any pair of points along a streamline within the flow. Let ρ be the density of the fluid that's flowing. Label the points we want to compare as Point 1 and Point 2. Choose a horizontal reference level, and let y_1 and y_2 be the heights of these points above this level. If the pressures at Points 1 and 2 are P_1 and P_2, and if the flow speeds at these points are v_1 and v_2, then Bernoulli's Equation says

Conservation of Energy
This equations looks very similar to a previous equation dealing with Conservation of Energy with Total Mechanical Energy. In fact, Bernoulli's Equation deals with the Conservation of Energy for fluids.

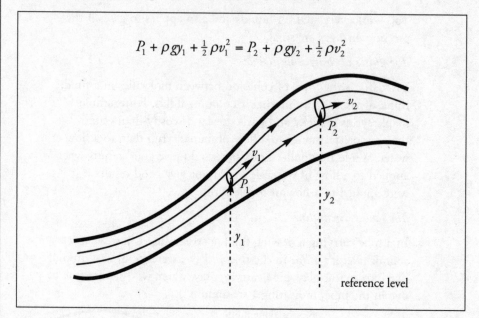

$$P_1 + \rho g y_1 + \tfrac{1}{2} \rho v_1^2 = P_2 + \rho g y_2 + \tfrac{1}{2} \rho v_2^2$$

An alternative, but equivalent, way of stating Bernoulli's Equation is to say that the quantity

$$P + \rho g y + \tfrac{1}{2} \rho v^2$$

is constant along a streamline. We mentioned earlier that Bernoulli's Equation is a statement of conversation of energy. You should notice the similarity between $\rho g y$ and mgh (gravitational potential energy) as well as between $\frac{1}{2} \rho v^2$ and $\frac{1}{2} m v^2$ (kinetic energy).

Torricelli's Theorem

Imagine that we punch a small hole in the side of an open tank of liquid. We can use Bernoulli's Equation to figure out how fast the liquid will flow out of the hole.

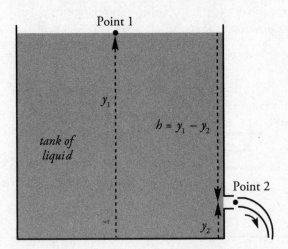

Let the bottom of the tank be our horizontal reference level, and choose Point 1 to be at the surface of the liquid and Point 2 to be at the hole where the water shoots out. First, the pressure at Point 1 is atmospheric pressure since the tank is open; and the emerging stream at Point 2 is open to the air, so it's at atmospheric pressure too. Therefore, $P_1 = P_2$, and these terms cancel out of Bernoulli's Equation. Next, since the area at Point 1 is so much greater than at Point 2, we can assume that v_1, the speed at which the water level in the tank drops, is much lower than v_2, the speed at which the water shoots out of the hole. (Remember that by the Continuity Equation, $A_1 v_1 = A_2 v_2$; since $A_1 \gg A_2$, we'll have $v_1 \ll v_2$.) Because $v_1 \ll v_2$, we can say that the water level at the top of the tank is approximately constant, and the speed of the water at the top is zero. So, Bernoulli's Equation becomes

$$\rho g y_1 = \rho g y_2 + \tfrac{1}{2} \rho v_2^2$$

Solving for v_2, we get

$$v_2 = \sqrt{2g(y_1 - y_2)} = \sqrt{2gh}$$

This is called Torricelli's Theorem.

Example 6 In the figure below, a pump forces water at a constant flow rate through a pipe whose cross-sectional area, A, gradually decreases: at the exit point, A has decreased to 1/3 its value at the beginning of the pipe. If $y = 60$ cm and the flow speed of the water just after it leaves the pump (Point 1 in the figure) is 1 m/s, what is the gauge pressure at Point 1?

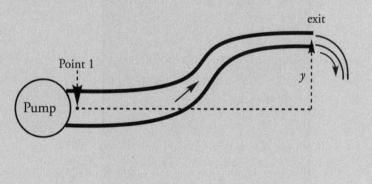

Solution. We'll apply Bernoulli's Equation to Point 1 and the exit point, Point 2. We'll choose the level of Point 1 as the horizontal reference level; this makes $y_1 = 0$. Now, because the cross-sectional area of the pipe decreases by a factor of 3 between Points 1 and 2, the flow speed must increase by a factor of 3; that is, $v_2 = 3v_1$. Since the pressure at Point 2 is p_{atm}, Bernoulli's Equation becomes

$$P_1 + \tfrac{1}{2}\rho v_1^2 = P_{atm} + \rho g y_2 + \tfrac{1}{2}\rho v_2^2$$

Now, P_1 is the total pressure at point 1. Recall that gauge pressure is $P_{tot} - P_{atm}$, so $P_{gauge} = P_1 - P_{atm}$.

Therefore,

$$
\begin{aligned}
P_1 - P_{atm} &= \rho g y_2 + \tfrac{1}{2}\rho v_2^2 - \tfrac{1}{2}\rho v_1^2 \\
&= \rho g y_2 + \tfrac{1}{2}\rho(3v_1)^2 - \tfrac{1}{2}\rho v_1^2 \\
&= \rho(g y_2 + 4v_1^2) \\
&= (1000 \text{ kg/m}^3)[(10 \text{ m/s}^2)(0.6 \text{ m}) + 4(1 \text{ m/s})^2] \\
&= 10^4 \text{ Pa}
\end{aligned}
$$

Example 7 The side of an above-ground pool is punctured, and water gushes out through the hole. If the total depth of the pool is 2.5 m, and the puncture is 1 m above ground level, what is the efflux speed of the water?

Solution. Torricelli's Theorem is $v = \sqrt{2gh}$, where h is the distance from the surface of the pool down to the hole. If the puncture is 1 m above ground level, then it's 2.5 − 1 = 1.5 m below the surface of the water (because the pool is 2.5 m deep). Therefore, the efflux speed will be

$$v = \sqrt{2gh} = \sqrt{2(10 \text{ m/s}^2)(1.5 \text{ m})} = \sqrt{30 \text{ m}^2/\text{s}^2} \approx 5.5 \text{ m/s}$$

The Bernoulli Effect

Consider the two points labeled in the pipe shown below:

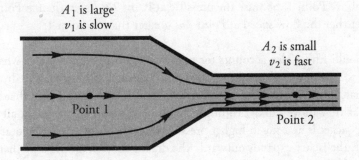

A_1 is large
v_1 is slow

A_2 is small
v_2 is fast

Point 1

Point 2

Since the heights y_1 and y_2 are equal in this case, the terms in Bernoulli's Equation that involve the heights will cancel, leaving us with

$$P_1 + \tfrac{1}{2}\rho v_1^2 = P_2 + \tfrac{1}{2}\rho v_2^2$$

We already know from the Continuity Equation $(f = Av)$ that the speed increases as the cross-sectional area of the pipe decreases; that is, since $A_2 < A_1$, we know that $v_2 > v_1$, so the equation above tells us that $P_2 < P_1$. This shows that

At comparable heights, the pressure is lower where the flow speed is greater.

This is known as the **Bernoulli** (or **Venturi**) **effect**, and is illustrated in the figure below.

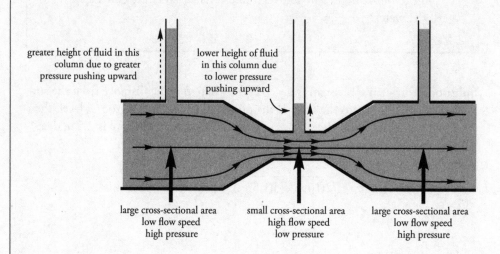

greater height of fluid in this column due to greater pressure pushing upward

lower height of fluid in this column due to lower pressure pushing upward

large cross-sectional area
low flow speed
high pressure

small cross-sectional area
high flow speed
low pressure

large cross-sectional area
low flow speed
high pressure

The height of the liquid column above Point 2 is less than the height of the liquid column above Point 1, because the pressure at Point 2 is lower than at Point 1, due to the fact that the flow speed at Point 2 is greater than at Point 1.

The Bernoulli Effect also accounts for many everyday phenomena. It's what allows airplanes to fly, curve balls to curve, and tennis balls hit with top spin to drop quickly. You may have seen sky divers or motorcycle riders wearing a jacket that seems to puff out as they move rapidly through the air. The essentially stagnant air trapped inside the jacket is at a much higher pressure than the air whizzing by outside, and as a result, the jacket expands outward. The drastic drop in air pressure that accompanies the high winds in a hurricane or tornado is yet another example. In fact, if high winds streak across the roof of a house whose windows are closed, the outside air pressure can be reduced so much that the air pressure inside the house (where the air speed is essentially zero) can be great enough to blow the roof off.

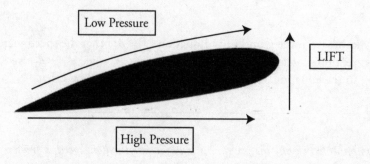

Low Pressure

LIFT

High Pressure

Air flow: The upper side of the wing is longer so air takes more time to reach the edge. If it takes same time (but longer distance) $[\uparrow d = \uparrow v\, t]$, then the top air is moving faster than bottom. The air on bottom has greater pressure and pushes up on the wing, giving airplanes lift force. .

Chapter 3 Review Questions

Solutions can be found in Chapter 12.

Section I: Multiple Choice

1. A large tank is filled with water to a depth of 6 m. If Point X is 1 m from the bottom and Point Y is 2 m from the bottom, how does p_X, the hydrostatic pressure due to the water at Point X, compare to p_Y, the hydrostatic pressure due to the water at Point Y ?

   (A) $p_X = 2p_Y$
   (B) $2p_X = p_Y$
   (C) $5p_X = 4p_Y$
   (D) $4p_X = 5p_Y$

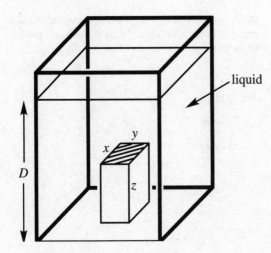

2. In the figure above, a box of dimensions x, y, and z rests on the bottom of a tank filled to depth D with a liquid of density ρ. If the tank is open to the atmosphere, what is the force on the (shaded) top of the box?

   (A) $xy(p_{atm} + \rho gD)$

   (B) $xyz[p_{atm} - \rho g(z - D)]$

   (C) $xy[p_{atm} + \rho g(D - z)]$

   (D) $xyz[p_{atm} + \rho gD]$

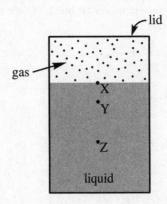

3. The figure above shows a closed container partially filled with liquid. Point Y is at a depth of 1 m, and Point Z is at a depth of 3 m. If the pressure at Point Y is 13,000 Pa, and the pressure at Point Z is 29,000 Pa, what is the pressure at the surface, Point X ?

   (A) 3,000 Pa
   (B) 4,000 Pa
   (C) 5,000 Pa
   (D) 7,000 Pa

4. A plastic cube 0.5 m on each side and with a mass of 100 kg floats in water. What fraction of the cube's volume is above the surface of the water?

   (A) $\dfrac{1}{5}$

   (B) $\dfrac{1}{4}$

   (C) $\dfrac{3}{4}$

   (D) $\dfrac{4}{5}$

5. A block of Styrofoam, with a density of ρ_S and volume V, is pushed completely beneath the surface of a liquid whose density is ρ_L, and released from rest. Given that $\rho_L > \rho_S$, which of the following expressions gives the magnitude of the block's initial upward acceleration?

(A) $(\rho_L - \rho_S)g$

(B) $\left(\dfrac{\rho_L}{\rho_S} - 1\right)g$

(C) $\left(\dfrac{\rho_L}{\rho_S} + 1\right)g$

(D) $\left[\left(\dfrac{\rho_L}{\rho_S}\right)^2 - 1\right]g$

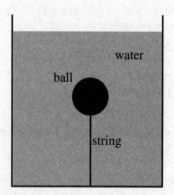

6. In the figure above, a ball with a density of 400 kg/m³ and volume 5×10^{-3} m³ is attached to a string, the other end of which is fastened to the bottom of the tank. If the tank is filled with water, what is the tension in the string?

(A) 20 N
(B) 30 N
(C) 40 N
(D) 50 N

7. An object with a density of 2000 kg/m³ weighs 100 N less when it's weighed while completely submerged in water than when it's weighed in air. What is the actual weight of this object?

(A) 200 N
(B) 300 N
(C) 400 N
(D) 600 N

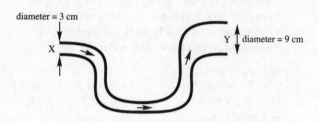

8. In the pipe shown above, which carries water, the flow speed at Point X is 6 m/s. What is the flow speed at Point Y ?

(A) $\dfrac{2}{3}$ m/s

(B) 2 m/s

(C) 18 m/s

(D) 54 m/s

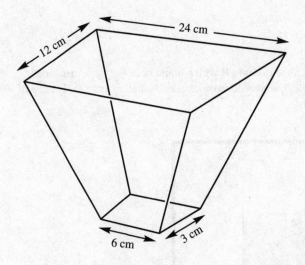

24 cm

12 cm

6 cm

3 cm

10. A pump is used to send water through a hose, the diameter of which is 10 times that of the nozzle through which the water exits. If the nozzle is 1 m higher than the pump, and the water flows through the hose at 0.4 m/s, what is the difference in pressure between the pump and the atmosphere?

(A) 108 kPa
(B) 260 kPa
(C) 400 kPa
(D) 810 kPa

9. The figure above shows a portion of a conduit for water, one with rectangular cross sections. If the flow speed at the top is v, what is the flow speed at the bottom?

(A) $4v$
(B) $8v$
(C) $12v$
(D) $16v$

Section II: Free Response

1. The figure below shows a tank open to the atmosphere and filled to depth D with a liquid of density ρ_L. Suspended from a string is a block of density ρ_B (which is greater than ρ_L), whose dimensions are x, y, and z (meters). The top of the block is at depth h meters below the surface of the liquid.

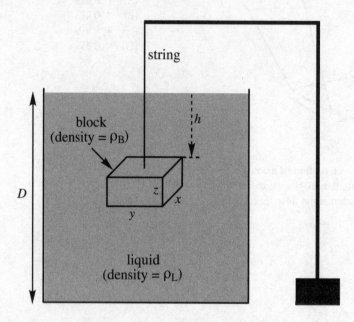

In each of the following, write your answer in simplest form in terms of ρ_L, ρ_B, x, y, z, h, D, and g.

(a) Find the force due to the pressure on the top surface of the block and on the bottom surface. Sketch these forces in the diagram below:

(b) What are the average forces due to the pressure on the other four sides of the block? Sketch these forces in the diagram above.

(c) What is the total force on the block due to the pressure?

(d) Find an expression for the buoyant force on the block. How does your answer here compare to your answer to part (c)?

(e) What is the tension in the string?

2. The figure below shows a large, cylindrical tank of water, open to the atmosphere, filled with water to depth D. The radius of the tank is R. At a depth h below the surface, a small circular hole of radius r is punctured in the side of the tank, and the point where the emerging stream strikes the level ground is labeled X.

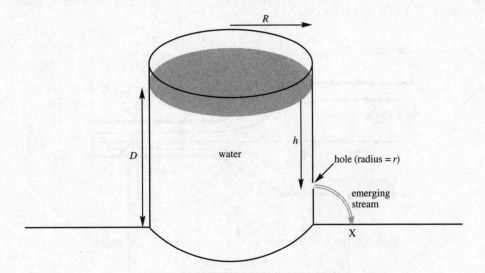

In parts (a) through (c), assume that the speed with which the water level in the tank drops is negligible.

(a)   At what speed does the water emerge from the hole?

(b)   How far is point X from the edge of the tank?

(c)   Assume that a second small hole is punctured in the side of the tank, a distance of $h/2$ directly above the hole shown in the figure. If the stream of water emerging from this second hole also lands at Point X, find h in terms of D.

(d)   For this part, do *not* assume that the speed with which the water level in the tank drops is negligible, and derive an expression for the speed of efflux from the hole punctured at depth h below the surface of the water. Write your answer in terms of $r, R, h,$ and g.

3. The figure below shows a pipe fitted with a Venturi U-tube. Fluid of density ρ_F flows at a constant flow rate and with negligible viscosity through the pipe, which constricts from a cross-sectional area A_1 at Point 1 to a smaller cross-sectional area A_2 at Point 2. The upper portion of both sides of the Venturi U-tube contain the same fluid that's flowing through the pipe, while the lower portion is filled with a fluid of density ρ_V (which is greater than ρ_F). At Point 1 in the pipe, the pressure is P_1 and the flow speed is v_1 ; at Point 2 in the pipe, the pressure is P_2 and the flow speed is v_2. All the fluid within the Venturi U-tube is stationary.

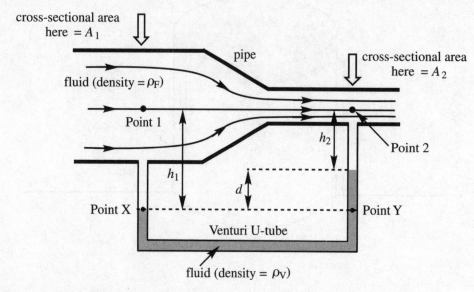

(a)   What is P_X, the hydrostatic pressure at Point X? Write your answer in terms of P_1, ρ_F, h_1, and g.

(b)   What is P_Y, the hydrostatic pressure at Point Y? Write your answer in terms of P_2, ρ_F, ρ_V , h_2, d, and g.

(c)   Write down the result of Bernoulli's Equation applied to Points 1 and 2 in the pipe, and solve for $P_1 - P_2$.

(d)   Since $P_X = P_Y$, set the expressions you derived in parts (a) and (b) equal to each other, and use this equation to find $P_1 - P_2$.

(e)   Derive an expression for the flow speed, v_2, and the flow rate, f, in terms of A_1, A_2, d, ρ_F, ρ_V, and g. Show that v_2 and f are proportional to $\sqrt{d}$.

Summary

o   Density is given by $\rho = \dfrac{m}{V}$. Pressure is given by $P = \dfrac{F}{A}$. Hydrostatic pressure can be found using $P = P_0 + \rho gh$ where ρgh is the pressure at a given depth below the surface of the fluid and P_0 is the pressure right above the surface of the fluid.

o   The buoyant force is an upward force any object immersed in a fluid experiences due to the displaced fluid. The buoyant force is given by $F_{buoy} = \rho Vg$, where ρ is the density of the fluid and V is the volume of the fluid displaced.

o   The Continuity Equation says that the flow rate through a pipe (area times velocity) is constant so that $A_1 v_1 = A_2 v_2$. This expresses the idea that a larger cross sectional area of pipe will experience fluids traveling at a lower velocity.

o   Bernoulli's Equation is a statement of conservation of energy. It states

$$\mathbf{P} + \rho \mathbf{gy} + \frac{1}{2}\rho \mathbf{v}^2 = \mathbf{constant}$$

126

Chapter 4
Thermal Physics

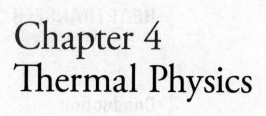

INTRODUCTION

This chapter looks at heat and temperature, concepts that seem familiar from our everyday experience. Technically, **heat** is defined as thermal energy transmitted from one body to another. While an object can contain **thermal energy** (due to the random motion of its molecules), an object doesn't *contain* heat; heat is energy *in transit*. **Temperature**, on the other hand, is a measure of an object's internal thermal energy.

HEAT TRANSFER

There are three principal modes by which energy can be transferred:

Conduction

An iron skillet is sitting on a hot stove, and you accidentally touch the handle. You notice right away that there's been a transfer of thermal energy to your hand. The process by which this happens is known as conduction. The highly agitated atoms in the handle of the hot skillet bump into the atoms of your hand, making them vibrate more rapidly, thus heating up your hand.

Convection

As the air around a candle flame warms, it expands, becomes less dense than the surrounding cooler air, and thus rises due to buoyancy. As a result, heat is transferred away from the flame by the large scale (from the atoms' point of view anyway) motion of a fluid (in the case, air). This is convection.

Radiation

Sunlight on your face warms your skin. Radiant energy from the sun's fusion reactions is transferred across millions of kilometers of essentially empty space via electromagnetic waves. Absorption of the energy carried by these light waves defines heat transfer by radiation.

Heat conducts from one point to another only if there is a temperature difference between the two objects. The rate at which heat is transferred is given by

$$\frac{Q}{\Delta t} = \frac{kA\Delta T}{L}$$

THE KINETIC THEORY OF GASES

Unlike the condensed phases of matter—solid and liquid—the atoms or molecules that make up a gas do not move around relatively fixed positions. Rather, the molecules of a gas move freely and rapidly, in a chaotic swarm. A confined gas exerts a force on the walls of its container, because the molecules are zipping around inside the container, striking the walls and rebounding. The magnitude of the the force per unit area is just the pressure, P, as defined in the previous chapter.

As we'll see, the faster the gas molecules are moving, the more pressure they exert.

THE IDEAL GAS LAW

Three physical properties—pressure (P), volume (V), and temperature (T)—describe a gas. At low densities, all gases approach *ideal* behavior; this means that these three variables are related by the equation

$$PV = nRT$$

Equation Sheet

where n is the number of moles of gas and R is a constant (8.31 J/mol·K) called the **universal gas constant**. This equation is known as the **Ideal Gas Law**. It can also be written as $PV = Nk_BT$, where k_B is Boltzmann's constant ($k_B = 1.38 \times 10^{-23}$ J/K). The ideal gas law tells us how these variables are related to one another for a gas with no intermolecular forces. This is a bad assumption for a charged gas, where the individual charges would interact electrically (a strong interaction), but it is a good assumption for a neutral gas, where the atoms would only interact gravitationally (a very weak interaction).

An important consequence of this equation is that, for a fixed volume of gas, an increase in P gives a proportional increase in T. The pressure increases when the gas molecules strike the walls of their container with more force, which occurs if they move more rapidly. Using Newton's Second Law (*rate of change of momentum = force*) we can find that the pressure exerted by N molecules of gas in a container of volume V is related to the average kinetic energy of the molecules by the equation $PV = \frac{2}{3} NK_{\text{avg}}$. Comparing this to the Ideal Gas Law we see that $\frac{2}{3} NK_{\text{avg}} = nRT$. We can rewrite this equation in the form $\frac{2}{3} N_A K_{\text{avg}} = RT$, since, by definition,

$N = nN_A$. The ratio R/N_A is a fundamental constant of nature called **Boltzmann's constant** ($k_B = 1.38 \times 10^{-23}$ J/K), so our equation becomes

$$K_{avg} = \frac{3}{2}k_B T$$

This tells us that the average translational kinetic energy of the gas molecules is directly proportional to the absolute temperature of the sample. Remember, this means you must use kelvins as your temperature unit.

Since the average kinetic energy of the gas molecules is $K_{avg} = \frac{1}{2}m(v^2)_{avg}$, the equation above becomes $\frac{1}{2}m(v^2)_{avg} = \frac{3}{2}k_B T$, so

$$\sqrt{(v^2)_{avg}} = \sqrt{\frac{3k_B T}{m}}$$

The quantity on the left-hand side of this equation, the square root of the average of the square of v, is called the **root-mean-square speed**, v_{rms}, so

$$v_{rms} = \sqrt{\frac{3k_B T}{m}}$$

It's important to realize that the molecules in the container have a wide range of speeds; some are much slower and others are much faster than v_{rms}. The root-mean-square speed is important because it gives us a type of average speed that's easy to calculate from the temperature of the gas.

Example 1 In order for the rms speed of the molecules in a given sample of gas to double, what must happen to the temperature?

Solution. Temperature is a measure of the average kinetic energy. The velocity is determined from the following equation:

$$v_{rms} = \sqrt{\frac{3k_b T}{m}}$$

Since v_{rms} is proportional to the square root of T, the temperature must quadruple, again, assuming the temperature is given in kelvins.

> **Example 2** A cylindrical container of radius 15 cm and height 30 cm contains 0.6 mole of gas at 433 K. How much force does the confined gas exert on the lid of the container?

Solution. The volume of the cylinder is $\pi r^2 h$, where r is the radius and h is the height. Since we know V and T, we can use the Ideal Gas Law to find P. Because pressure is force per unit area, we can find the force on the lid by multiplying the gas pressure times the area of the lid.

$$P = \frac{nRT}{V} = \frac{(0.6 \text{ mol})(8.31 \text{ J/mol} \cdot \text{K})(433 \text{ K})}{\pi (0.15 \text{ m})^2 (0.30 \text{ m})} = 1.018 \times 10^5 \text{ Pa}$$

So, since the area of the lid is πr^2, the force exerted by the confined gas on the lid is:

$$F = PA = (1.018 \times 10^5 \text{ Pa}) \cdot \pi (0.15 \text{ m})^2 = 7,200 \text{ N}$$

This is about 1,600 pounds of force, which seems like a lot. Why doesn't this pressure pop the lid off? Because, while the bottom of the lid is feeling a pressure (due to the confined gas) of 1.018×10^5 Pa that exerts a force upward, the top of the lid feels a pressure of 1.013×10^5 Pa (due to the atmosphere) that exerts a force downward. The net force on the lid is

$$F_{net} = (\Delta P)A = (0.005 \times 10^5 \text{ Pa}) \cdot \pi (0.15 \text{ m})^2 = 35 \text{ N}$$

which is only about 8 pounds.

THE LAWS OF THERMODYNAMICS

We've learned about two ways in which energy may be transferred between a system and its environment. One is work, which takes place when a force acts over a distance. The other is heat, which takes place when energy is transferred due to a difference in temperature. The study of the energy transfers involving work and heat, and the resulting changes in internal energy, temperature, volume, and pressure is called **thermodynamics**.

The Zeroth Law of Thermodynamics

When two objects are brought into contact, heat will flow from the warmer object to the cooler one until they reach thermal equilibrium. This property of temperature is expressed by the Zeroth Law of Thermodynamics:

> If Objects 1 and 2 are each in thermal equilibrium with Object 3, then Objects 1 and 2 are in thermal equilibrium with each other.

The First Law of Thermodynamics

Simply put, the First Law of Thermodynamics is a statement of the Conservation of Energy that includes heat.

> Energy (in the form of heat) is neither created nor destroyed in any thermodynamic system.

Consider the following example, which is the prototype that's studied extensively in thermodynamics.

An insulated container filled with an ideal gas rests on a heat reservoir (that is, something that can act as a heat source or a heat sink). The container is fitted with a snug, but frictionless, weighted piston that can be raised or lowered. The confined gas is the *system*, and the piston and heat reservoir are the *surroundings*.

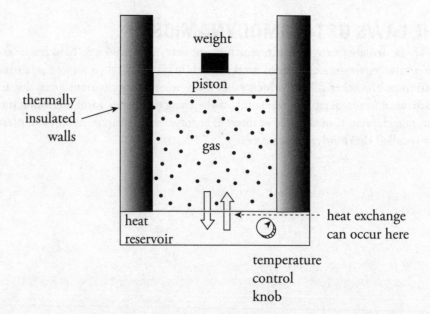

The **state** of the gas is given once its pressure, volume, and temperature are known, and the equation that connects these state variables is the Ideal Gas Law, $PV = nRT$. We'll imagine performing different experiments with the gas, such as heating it or allowing it to cool or increasing or decreasing the weight on the piston, and study the energy transfers (work and heat) and the changes in the state variables. If each process is carried out such that, at each moment, the system and its surroundings are in thermal equilibrium, we can plot the pressure (P) versus the volume (V) on a diagram. By following the path of this **P–V diagram**, we can study how the system is affected as it moves from one state to another.

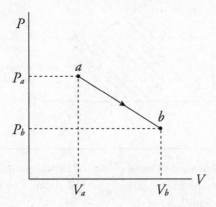

Work is done on or by the system when the piston is moved and the volume of the gas changes. For example, imagine that the weight pushes the piston downward a distance d, causing a decrease in volume. Assume that the pressure stays constant at P. (Heat must be removed via the reservoir to accomplish this.) We can calculate the work done on the gas during this compression as $W = -Fd$, but since $F = PA$, we have $W = -PAd$, and because $Ad = \Delta V$, we have

$$W = -P\Delta V$$

Equation Sheet

Textbooks differ about the circumstances under which work in thermodynamics is defined to be positive or negative. The negative signs we have included in the equations above are consistent with those used in the AP Physics Exam (appearing in the equation sheet given for Section II). For the exam, work in thermodynamics is considered to be positive when the work is being done *on the system*. This means that the volume of the system is *decreasing*, ΔV is *negative*, and, in agreement with intuition, energy is being *added* to the system. In other words, when work is done in compressing a system, ΔV is *negative* and the work done, $W = -P\Delta V$, is *positive*. This also means, conversely, that when the system is doing work *on the surroundings* (volume *increasing*, ΔV *positive*), the work is *negative*, in agreement with intuition that energy is leaving the system.

The equation $W = -P\Delta V$ assumes that the pressure P does not change during the process. If P *does* change, then the work is equal to the area under the curve in the P–V diagram; moving left to right as the volume increases gives a negative area (and negative work), while moving right to left as the volume decreases gives a positive area (and positive work). Because of this difference when moving left to right or right to left, processes in P–V diagrams are drawn with arrows to indicate which direction the process follows. In the diagram below, the process is taken from an initial state *a*, with a particular *P* and *V* value to a final state *b*, where *P* and *V* have changed.

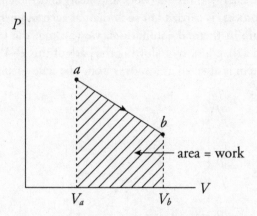

Example 3 What's the value of *W* for the process *ab* following path 1 and for the same process following path 2 (from *a* to *d* to *b*), shown in the P–V diagram below?

Solution.

Path 1. Since, in path 1, *P* remains constant, the work done is just –*P* Δ *V*:

$$W = -P\Delta V = -(1.5 \times 10^5 \text{ Pa})[(30 \times 10^{-3} \text{ m}^3) - (10 \times 10^{-3} \text{ m}^3)] = -3{,}000 \text{ J}$$

Path 2. If the gas is brought from state *a* to state *b*, along path 2, then work is done only along the part from *a* to *d*. From *d* to *b*, the volume of the gas does not change, so no work can be performed. The area under the graph from *a* to *d* is

$$W = -\frac{1}{2}h(b_1 + b_2) = -\frac{1}{2}(\Delta V)(P_a + P_d)$$
$$= -\frac{1}{2}(20 \times 10^{-3} \text{ m}^3)[(1.5 \times 10^5 \text{ Pa}) + (0.7 \times 10^5 \text{ Pa})]$$
$$= -2{,}200 \text{ J}$$

As this example shows, the amount of work done by a gas depends not only on the initial and final states of the system, but also on the path between the two. In general, different paths give different values for W.

Experiments have shown that the value of $Q + W$ is *not* path dependent; it depends only on the initial and the final state of the system, so it describes a change in a fundamental property. This property is called the system's **internal energy**, denoted U, and the change in the system's internal energy, ΔU, is equal to $Q + W$. This is true regardless of the process that brought the system from its initial to final state. This statement is known as

Equation Sheet

The First Law of Thermodynamics

$$\Delta U = Q + W$$

ΔU depends only on temperature change

(If volume is constant, $W = 0$ and $U = Q$.)

This statement of the First Law is consistent with the interpretation of work ($W = -P\Delta V$) explained above. The First Law identifies W and Q as separate physical mechanisms for adding to or removing energy from the system, and the signs of both W and Q are defined consistently: Both are positive when they are adding energy to the system and negative when they are removing energy from the system.

Example 4 A 0.5 mol sample of an ideal gas is brought from state a to state b when 7500 J of heat is added along the path shown in the following P–V diagram:

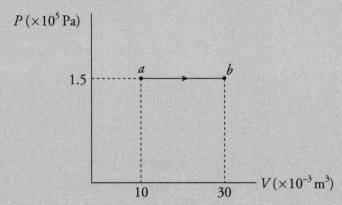

What are the values of each of the following?

    (a) the temperature at a

    (b) the temperature at b

    (c) the work done by the gas during process ab

    (d) the change in the internal energy of the gas

Solution.

(a, b)   Both of these questions can be answered using the Ideal Gas Law, $T = PV/(nR)$:

$$T_a = \frac{P_a V_a}{nR} = \frac{(1.5 \times 10^5 \text{ Pa})(10 \times 10^{-3} \text{ m}^3)}{(0.5 \text{ mol})(8.31 \text{ J/mol} \cdot \text{K})} = 360 \text{ K}$$

$$T_b = \frac{P_b V_b}{nR} = \frac{(1.5 \times 10^5 \text{ Pa})(30 \times 10^{-3} \text{ m}^3)}{(0.5 \text{ mol})(8.31 \text{ J/mol} \cdot \text{K})} = 1{,}080 \text{ K}$$

(c)   Since the pressure remains constant during the process, we can use the equation $W = -P\Delta V$. Because $\Delta V = (30 - 10) \times 10^{-3} \text{ m}^3 = 20 \times 10^{-3} \text{ m}^3$, we find that

$$W = -P\Delta V = (1.5 \times 10^5 \text{ Pa})(20 \times 10^{-3} \text{ m}^3) = -3{,}000 \text{ J}$$

The expanding gas did negative work against its surroundings, pushing the piston upward. Important note: If the pressure remains constant (which is designated by a horizontal line in the P–V diagram), the process is called **isobaric**.

(d)   By the First Law of Thermodynamics,

$$\Delta U = Q + W = 7{,}500 - 3{,}000 \text{ J} = 4{,}500 \text{ J}$$

Example 5 A 0.5 mol sample of an ideal monatomic gas is brought from state a to state b along the path shown in the following P–V diagram:

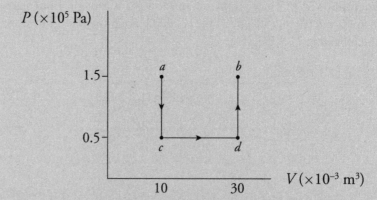

What are the values of each of the following?

(a)   the work done by the gas during process ab

(b)   the change in the internal energy of the gas

(c)   the heat added to the gas during process ab

Solution. Note that the initial and final states of the gas are the same as in the preceding example, but the path is different.

(a)   Let's break the path into 3 pieces:

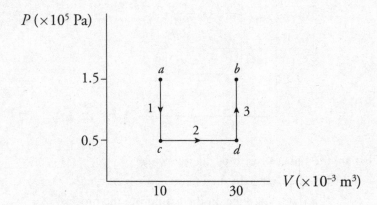

Over paths 1 and 3, the volume does not change, so no work is done. Work is done only over path 2:

$$W = -P\Delta V = -(0.5 \times 10^5 \text{ Pa})(20 \times 10^{-3} \text{ m}^3) = -1,000 \text{ J}$$

Once again, the expanding gas does negative work against its surroundings, pushing the piston upward.

(b)   Because the initial and final states of the gas are the same here as they were in the preceding example, the change in internal energy, ΔU, *must* be the same. Therefore, $\Delta U = 4,500$ J.

(c)   By the First Law of Thermodynamics, $\Delta U = Q + W$, so

$$Q = U - W = 4,500\text{J} - (-1,000 \text{ J}) = 5,500 \text{ J}$$

Example 6 An **isochoric** process is one that takes place with no change in volume. What can you say about the change in the internal energy of a gas if it undergoes an isochoric change of state?

Solution. An isochoric process is illustrated by a vertical line in a *P–V* diagram and, since no change in volume occurs, $W = 0$. By the First Law of Thermodynamics, $\Delta U = Q + W = Q$. Therefore, the change in internal energy is entirely due to (and equal to) the heat transferred. If heat is transferred into the system (positive Q), then ΔU is positive; if heat is transferred out of the system (negative Q), then ΔU is negative.

Example 7 A 0.5 mol sample of an ideal gas is brought from state a back to state a along the path shown in the following P–V diagram:

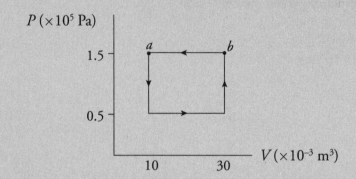

What are the values of each of the following?

      (a) the change in the internal energy of the gas
      (b) the work done on the gas during the process
      (c) the heat added to the gas during the process

Solution. A process such as this, which begins and ends at the same state, is said to be cyclical.

(a)   Because the final state is the same as the initial state, the internal energy of the system cannot have changed, so $\Delta U = 0$.

(b)   The total work involved in the process is equal to the work done from c to d plus the work done from b to a,

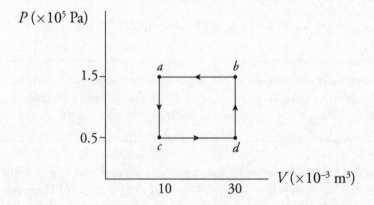

because only along these paths does the volume change. Along these portions, we find that

$$W_{cd} = -P\Delta V_{cd} = (0.5 \times 10^5 \text{ Pa})(+20 \times 10^{-3} \text{ m}^3) = -1{,}000 \text{ J}$$

$$W_{cd} = -P\Delta V_{ba} = (1.5 \times 10^5 \text{ Pa})(-20 \times 10^{-3} \text{ m}^3) = +3{,}000 \text{ J}$$

so the total work done is $W = +2{,}000$ J. The fact that W is positive means that, overall, work was done *on* the gas by the surroundings. Notice that for a cyclical process, the total work done is equal to the area enclosed by the loop, with clockwise travel taken as negative and counterclockwise travel taken as positive.

(c) The First Law of Thermodynamics states that $\Delta U = Q + W$. Since $\Delta U = 0$, it must be true that $Q = -W$ (which will always be the case for a cyclical process), so $Q = -2{,}000$ J.

Example 8 A 0.5 mol sample of an ideal gas is brought from state a to state d along an **isotherm**, then isobarically to state c and isochorically back to state a, as shown in the following P–V diagram:

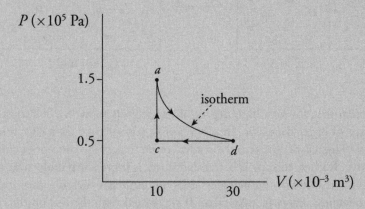

A process that takes place with no variation in temperature is said to be **isothermal**. Given that the work done during the isothermal part of the cycle is $-1{,}650$ J, how much heat is transferred during the isothermal process, from a to d?

Solution. Be careful that you don't confuse *isothermal* with *adiabatic*. A process is isothermal if the *temperature* remains constant; a process is **adiabatic** if $Q = 0$. You might ask, *How could a process be isothermal without also being adiabatic at the same time?* Remember that the temperature is determined by the internal energy of the gas, which is affected by changes in Q, W, or both. Therefore, it's possible for U to remain unchanged even if Q is not 0 (because there can be an equal but opposite W to cancel it out). In fact, this is the key to this problem. Since T doesn't change from a to d, neither can the internal energy, which depends entirely on T. Because $\Delta U_{ad} = 0$, it must be true that $Q_{ad} = -W_{ad}$. Since W_{ad} equals $-1{,}650$ J, Q_{ad} must be $+1{,}650$ J. The gas absorbs heat from the reservoir and uses all this energy to do negative work as it expands, pushing the piston upward.

The Second Law of Thermodynamics

Entropy

Consider a box containing two pure gases separated by a partition. What would happen if the partition were removed? The gases would mix, and the positions of the gas molecules would be random.

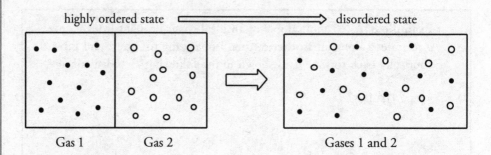

A closed system that shows a high degree of order tends to evolve in such a way that its degree of order decreases. In other words, disorder (or, as it's technically called, entropy) increases. The term "disorder" often makes the concept of entropy sound negative. Rather, entropy is better described as **increasing molecular freedom**. This is the reason why broken glass does not put itself back together and the reason why chemical reactions take place. If we started with the box on the right, containing the mixture of the gases, it would be virtually impossible that at any later time all the molecules of Gas 1 would happen to move to the left side of the box and, at the same time, the molecules of Gas 2 would spontaneously move to the right side of the box. If we were to watch a movie of this process, and saw the mixed-up molecules suddenly separate and move to opposite sides of the box, we'd assume that the film was running backwards. In a way, the second law of thermodynamics defines the direction of time. Time flows in such a way that ordered systems become disordered. Disordered states do not spontaneously become ordered without any other changes taking place. The following is the essence of one form of the second law of thermodynamics:

> The total amount of disorder—the total entropy—of a system plus its surroundings will never decrease.

It is possible for the entropy of a system to decrease, but it will always be at the expense of a greater increase in entropy in the surroundings. For example, when water freezes, its entropy decreases. The molecules making up an ice crystal have a more structured order than the random collection of water molecules in the liquid phase, so the entropy of the water decreases when it freezes. But when water freezes, it releases heat energy into its environment, which creates disorder in the surroundings. If we were to figure out the total change in entropy of the water plus its surroundings, we would find that although the entropy of the water itself decreased, it was more than compensated by a greater amount of entropy increase in the surroundings. So, the total entropy of the system and its surroundings increased, in agreement with the second law of thermodynamics.

There are several equivalent statements of the second law. In addition to the entropy form says that heat always flows from an object at higher temperature to an object at lower temperature, never the other way around. Heat always flows from hot to cold, never cold to hot. Another form, which will be considered more fully in a moment, says that a heat engine can never operate at 100% efficiency, or equivalently, that it is impossible to convert heat completely into work.

Heat Engines

The form of the Second Law of Thermodynamics that we'll study for the AP Physics Exam deals with heat engines.

Converting work to heat is easy—rubbing your hands together in order to warm them up shows that work can be completely converted to heat. What we'll look at it is the reverse process: How efficiently can heat be converted into work? A device that uses heat to produce useful work is called a **heat engine**. The internal-combustion engine in a car is an example. In particular, we're only interested in engines that take their working substance (a mixture of air and fuel in this case) through a cyclic process, so that the cycle can be repeated. The basic components of any cyclic heat engine are simple: Energy in the form of heat comes into the engine from a high-temperature source, some of this energy is converted into useful work, the remainder is ejected as exhaust heat into a low-temperature sink, and the system returns to its original state to run through the cycle again.

Entropy and Heat
A system always increases in entropy over time. This entropy is usually in the form of heat given off.

All engines function to convert heat into work by exchanging energy from a reservoir at higher temperature to reservoirs of lower temperature. A good example of this is a refrigerator. In a fridge, a liquid moves through tubes to remove heat from the internal chamber (such as the freezer) to the outer air. Thus, cold is not actually a concept, but the sensation of low heat content. This is why the back of a refrigerator feels hot.

Since we're looking at cyclic engines only, the system returns to its original state at the end of each cycle, so ΔU must be 0. Therefore, by the First Law of Thermodynamics, $Q_{net} = -W$. That is, the net heat absorbed by the system is equal to the work performed by the system. The heat absorbed from the high-temperature source is denoted Q_H (H for *hot*), and the heat that is discharged into the low-temperature reservoir is denoted Q_C (C for *cold*). Because heat coming *in* is positive and heat going *out* is negative, Q_H is positive and Q_C is negative, and the net heat absorbed is $Q_H + Q_C$. Instead of writing Q_{net} in this way, it's customary to write it as $Q_H - |Q_C|$, to show explicitly that Q_{net} is less than Q_H.

This is one of the forms of

The Second Law of Thermodynamics
For any cyclic heat engine, some exhaust heat is always produced.
It's impossible to completely convert heat into useful work.

Example 9 A heat engine draws 800 J of heat from its high-temperature source and discards 450 J of exhaust heat into its cold-temperature reservoir during each cycle. How much work does this engine perform per cycle?

Solution. The absolute value of the work output per cycle is equal to the difference between the heat energy drawn in and the heat energy discarded:

$$|W| = Q_H - |Q_C| = 800 \text{ J} - 450 \text{ J} = 350 \text{ J}$$

Chapter 4 Review Questions

Solutions can be found in Chapter 12.

Section I: Multiple Choice

1. A container holds a mixture of two gases, CO_2 and H_2, in thermal equilibrium. Let K_C and K_H denote the average kinetic energy of a CO_2 molecule and an H_2 molecule, respectively. Given that a molecule of CO_2 has 22 times the mass of a molecule of H_2, the ratio K_C / K_H is equal to

   (A) 1/22
   (B) 1
   (C) $\sqrt{22}$
   (D) 22

2. If the temperature and volume of a sample of an ideal gas are both doubled, then the pressure

   (A) decreases by a factor of 2
   (B) increases by a factor of 2
   (C) increases by a factor of 4
   (D) remains unchanged

3. In three separate experiments, a gas is transformed from state P_i, V_i to state P_f, V_f along the paths (1, 2, and 3) illustrated in the figure below:

   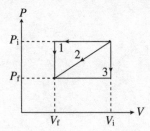

   The work done on the gas is

   (A) greatest for path 1
   (B) least for path 2
   (C) greatest for path 2
   (D) the same for all three paths

4. An ideal gas is compressed isothermally from 20 m³ to 10 m³. During this process, 5 J of work is done to compress the gas. What is the change of internal energy for this gas?

   (A) −10 J
   (B) −5 J
   (C) 0 J
   (D) 5 J

5. An ideal gas is confined to a container whose volume is fixed. If the container holds n moles of gas, by what factor will the pressure increase if the absolute temperature is increased by a factor of 2 ?

   (A) $2/(nR)$
   (B) 2
   (C) $2nR$
   (D) $2/n$

6. Two large glass containers of equal volume each hold 1 mole of gas. Container 1 is filled with hydrogen gas (2 g/mol), and Container 2 holds helium (4 g/mol). If the pressure of the gas in Container 1 equals the pressure of the gas in Container 2, which of the following is true?

   (A) The temperature of the gas in Container 1 is lower than the temperature of the gas in Container 2.
   (B) The temperature of the gas in Container 1 is greater than the temperature of the gas in Container 2.
   (C) The rms speed of the gas molecules in Container 1 is lower than the rms speed of the gas molecules in Container 2.
   (D) The rms speed of the gas molecules in Container 1 is greater than the rms speed of the gas molecules in Container 2.

7. Through a series of thermodynamic processes, the internal energy of a sample of confined gas is increased by 560 J. If the net amount of work done on the sample by its surroundings is 320 J, how much heat was transferred between the gas and its environment?

   (A) 240 J absorbed
   (B) 240 J dissipated
   (C) 880 J absorbed
   (D) 880 J dissipated

8. What's the total work performed on the gas as it's transformed from state *a* to state *c*, along the path indicated?

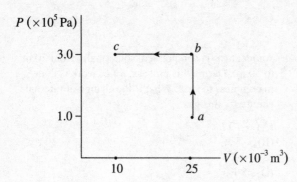

(A) 1,500 J
(B) 3,000 J
(C) 4,500 J
(D) 9,500 J

9. In one of the steps of the Carnot cycle, the gas undergoes an isothermal expansion. Which of the following statements is true concerning this step?

(A) No heat is exchanged between the gas and its surroundings, because the process is isothermal.
(B) The temperature decreases because the gas expands.
(C) The internal energy of the gas remains constant.
(D) The internal energy of the gas decreases due to the expansion.

10. A cup of hot coffee is sealed inside a perfectly thermally insulating container. A long time is allowed to pass. Which of the following correctly explains the final thermal configuration within the box?

(A) The coffee has not changed temperature because the box is perfectly insulating.
(B) The coffee has gotten warmer and the air in the box has gotten cooler because of an exchange of thermal energy between the air and the coffee.
(C) The coffee has gotten cooler and the air in the box has gotten warmer because of an exchange of thermal energy between the air and the coffee.
(D) The coffee has gotten cooler but the air in the box has not changed its temperature. The energy from the coffee has caused an increase in entropy within the box.

Section II: Free Response

1. When a system is taken from state a to state b along the path acb shown in the figure below, 70 J of heat flows into the system, and the system does 30 J of work.

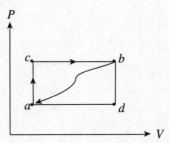

   (a) When the system is returned from state b to state a along the curved path shown, 60 J of heat flows out of the system. Does the system perform work on its surroundings or do the surroundings perform work on the system? How much work is done?

   (b) If the system does 10 J of work in transforming from state a to state b along path adb, does the system absorb or does it emit heat? How much heat is transferred?

   (c) If $U_a = 0$ J and $U_d = 30$ J, determine the heat absorbed in the processes db and ad.

   (d) For the process $adbca$, identify each of the following quantities as positive, negative, or zero:

   $W = $ _____ $Q = $ _____ $\Delta U = $ _____

2. A 0.4 mol sample of an ideal diatomic gas undergoes slow changes from state a to state b to state c and back to a along the cycle shown in the P–V diagram below:

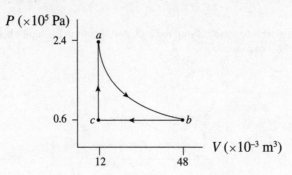

Path ab is an isotherm, and it can be shown that the work done by the gas as it changes isothermally from state a to state b is given by the equation

$$W_{ab} = -nRT \ \ln\frac{V_b}{V_a}$$

The molar heat capacities for the gas are C_V = 20.8 J/mol·K and C_P = 29.1 J/mol·K.

(a)   What's the temperature of

(i)   state a?

(ii)   state b?

(iii)   state c?

(b)   Determine the change in the internal energy of the gas for

(i)   step ab

(ii)   step bc

(iii)   step ca

(c)   How much work, W_{ab}, is done by the gas during step ab?

(d)   What is the total work done over cycle $abca$?

(e)   (i)   Is heat absorbed or discarded during step ab?

(ii)   If so, how much?

(f)   What is the maximum possible efficiency (without violating the Second Law of Thermodynamics) for a cyclical heat engine that operates between the temperatures of states a and c?

Summary

○ In an isolated system, $Q_{gained} + Q_{lost} = 0$.

○ For phase changes there is no temperature change and the equation that relates the amount of heat needed to change a certain mass to its new state is given by $Q = mL$.

○ The rate at which heat is transferred is given by $H = \dfrac{Q}{t}$ or $H = \dfrac{kA\Delta T}{L}$, where k is the thermal conductivity (a property of the material), A is the cross-sectional area, ΔT is the temperature difference between the two sides and L is the thickness or distance between the two ends of the material.

○ For gases there are a few common equations to understand:

  • Pressure is defined as the fore per unit area ($P = \dfrac{F}{A}$).

  • The ideal gas law is expressed as either $PV = nRT$ or $PV = Nk_BT$.

  • The average kinetic energy of the gas molecules is given by $K_{avg} = \dfrac{3}{2}k_bT$.

  • The average speed of a molecule of the gas is given by $v_{rms} = \sqrt{\dfrac{3RT}{M}}$ or $v_{rms} = \sqrt{\dfrac{k_BT}{\mu}}$.

○ The work done on a gas is given by $W = -P\Delta V$ or can be found by the area under a pressure-vs.-volume graph.

○ The First Law of Thermodynamics is $\Delta U = Q+W$ where ΔU depends only on the temperature change.

○ *W* positive (or negative) means energy is being added to (or subtracted from) the system by means of work done on the system by the surroundings (or on the surroundings by the system).

○ *Q* positive (or negative) means energy is being added to (or subtracted from) the system by means of a flow of heat from the higher temperature surroundings (or system) to the lower temperature system (or surroundings).

Note: Some textbooks define work in thermodynamics in a different way: Work is considered to be positive when work is done on the surroundings. This is consistent with the idea that the overall objective of a heat engine is to produce external (positive) work. Under this definition, the First Law must be written as $U = Q - W$ (or $U + W = Q$), and *W* must then be interpreted differently from *Q*. That is, while *Q* is still positive when heat is being added to the system, *W* is now positive when work is being done by the system on the surroundings (thus decreasing the internal energy of the system).

Chapter 5
Electronic Forces
and Fields

INTRODUCTION

Long before the knowledge of electricity many people were aware of the mysterious force. Descriptions of electricity was reported by Egyptians, Greeks, and Romans. Electricity would remain a mystery until 1600, when English scientist William Gilbert made a careful study of electricity and magnetism. Further work was continued by other notable scientists including Ben Franklin (his famous metal key and kite experiment in a storm). Real progress came in the nineteenth century, with electricity and magnetism, by Hans Christian Orsted, Andre-Marie Ampere, Michael Faraday, and Georg Ohm. James Clerk Maxwell would difinitively link them together into what we call today electromagnetism.

ELECTRIC CHARGE

The basic components of atoms are protons, neutrons, and electrons. Protons and neutrons form the nucleus (and are referred to collectively as *nucleons*), while the electrons keep their distance, swarming around the nucleus. Most of an atom consists of empty space. In fact, if a nucleus were the size of the period at the end of this sentence, then the electrons would be 5 meters away. So what holds such an apparently tenuous structure together? One of the most powerful forces in nature: the *electromagnetic force*. Protons and electrons have a quality called **electric charge** that gives them an attractive force. Electric charge comes in two varieties; positive and negative. A positive particle always attracts a negative particle, and particles of the same charge always repel each other. Protons are positively charged, and electrons are negatively charged.

Protons and electrons are intrinsically charged, but bulk matter is not. This is because the amount of charge on a proton exactly balances the charge on an electron, which is quite remarkable in light of the fact that protons and electrons are very different particles. Since most atoms contain an equal number of protons and electrons, their overall electric charge is 0, because the negative charges cancel out the positive charges. Therefore, in order for matter to be **charged**, an imbalance between the numbers of protons and electrons must exist. This can be accomplished by either the removal or addition of electrons (that is, by the **ionization** of some of the object's atoms). If you remove electrons, then the object becomes positively charged, while if you add electrons, then it becomes negatively charged. Furthermore, charge is **conserved**. For example, if you rub a glass rod with a piece of silk, then the silk will acquire a negative charge and the glass will be left with an *equal* positive charge. *Net charge cannot be created or destroyed.* (*Charge* can be created or destroyed—it happens all the time—but *net* charge cannot.)

The magnitude of charge on an electron (and therefore on a proton) is denoted e. This stands for **elementary charge**, because it's the basic unit of electric charge. The charge of an ionized atom must be a whole number times e, because charge can be added or subtracted only in lumps of size e. For this reason we say that charge is **quantized**. To remind us of the quantized nature of electric charge, the charge of a particle (or object) is denoted by the letter q. In the SI system of units, charge is expressed in **coulombs** (abbreviated **C**). One coulomb is a tremendous amount of charge, about 10^{18} electrons. The value of e is about 1.6×10^{-19} C.

COULOMB'S LAW

The electric force between two charged particles obeys a law that is very similar to that describing the gravitational force between two masses: they are both inverse-square laws. The **electric force** between two particles with charges of q_1 and q_2, separated by a distance r, is given by the equation

$$F_E = k \frac{q_1 q_2}{r^2}$$

This is **Coulomb's Law**. We interpret a negative F_E as an attraction between the charges and a positive F_E as a repulsion. The value of the proportionality constant, k, depends on the material between the charged particles. In empty space (vacuum)—or air, for all practical purposes—it is called **Coulomb's constant** and has the approximate value $k_0 = 9 \times 10^9$ N·m²/C². k_0 is sometimes written in terms of a fundamental constant known as the **permittivity of free space**, denoted ε_0:

$$k_0 = \frac{1}{4\pi\varepsilon_0}$$

> **Example 1** Consider the proton and electron in hydrogen. The proton has a mass of 1.6×10^{-27} kg and a charge of $+e$. The electron has a mass of 9.1×10^{-31} kg and a charge of $-e$. In hydrogen, they are separated by a Bohr radius, about 0.5×10^{-10} m. What is the electric force between the proton and electron in hydrogen? What about the gravitational force between the proton and electron?

Solution. The electric force between the proton and the electron is given by Coulomb's Law:

$$F_E = \frac{1}{4\pi\varepsilon_0} \frac{q_1 q_2}{r^2} = (9 \times 10^9 \text{ N·m}^2/\text{C}^2) \frac{(1.6 \times 10^{-19} \text{ C})(-1.6 \times 10^{-19} \text{ C})}{(0.5 \times 10^{-10} \text{ m})^2} = -9.24 \times 10^{-8} \text{ N}$$

The fact that F_E is negative means that the force is one of *attraction*, which we naturally expect, since one charge is positive and the other is negative. The force between the proton and electron is along the line that joins the charges, as we've illustrated below. The two forces shown form an action/reaction pair.

$$q_1 \oplus \xrightarrow{\quad \mathbf{F}_E \quad} \qquad \xleftarrow{\quad \mathbf{F}_E \quad} \ominus q_2$$

Now, the gravitational force between the two charges is given by Newton's law of gravitation:

$$\mathbf{F}_G = G \times \frac{(m_1 m_2)}{r^2} = (6.7 \times 10^{-11} \, \frac{\text{kgm}^2}{\text{kg}^2}) \times \frac{(1.67 \times 10^{-27} \, \text{kg})(9.11 \times 10^{-31} \, \text{kg})}{(0.5 \times 10^{-10} \, \text{m})^2} = 2.25 \times 10^{-47} \, \text{N}$$

Now, compare the orders of magnitude of the electric force to the gravitational force. The electric force has an order of magnitude of 10^{-8} N, but the gravitational force has an order of magnitude of 10^{-47} N! This means that the electric force is something like 10^{39} times larger!

Addition of Forces

Consider three point charges: q_1, q_2, and q_3. The total electric force acting on, say, q_2 is simply the sum of $\mathbf{F}_{1\text{-on-}2}$, the electric force on q_2 due to q_1, and $\mathbf{F}_{3\text{-on-}2}$, the electric force on q_2 due to q_3:

$$\mathbf{F}_{\text{on } 2} = \mathbf{F}_{1\text{-on-}2} + \mathbf{F}_{3\text{-on-}2}$$

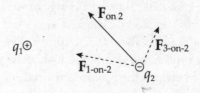

Example 2 Consider four equal, positive point charges that are situated at the vertices of a square. Find the net electric force on a negative point charge placed at the square's center.

Solution. Refer to the diagram below. The attractive forces due to the two charges on each diagonal cancel out: $F_1 + F_3 = 0$, and $F_2 + F_4 = 0$, because the distances between the negative charge and the positive charges are all the same and the positive charges are all equivalent. Therefore, by symmetry, the net force on the center charge is zero.

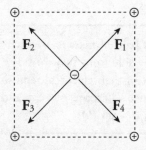

Example 3 If the two positive charges on the bottom side of the square in the previous example were removed, what would be the net electric force on the negative charge? Assume that each side of the square is 4.0 cm, each positive charge is 1.5 μC, and the negative charge is –6.2 nC. (μ is the symbol for "micro-," which equals 10^{-6}.)

Solution. If we break down F_1 and F_2 into horizontal and vertical components, then by symmetry the two horizontal components will cancel each other out, and the two vertical components will add:

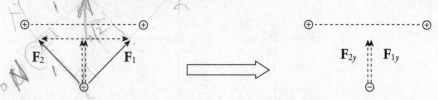

Since the diagram on the left shows the components of $\mathbf{F}_1$ and $\mathbf{F}_2$ making right triangles with legs each of length 2 cm, it must be that $F_{1y} = F_1 \sin 45°$ and $F_{2y} = F_2 \sin 45°$. Also, the magnitude of $\mathbf{F}_1$ equals that of $\mathbf{F}_2$. So the net electric force on the negative charge is $F_{1y} + F_{2y} = 2F \sin 45°$, where F is the strength of the force between the negative charge and each of the positive charges.

If s is the length of each side of the square, then the distance r between each positive charge and the negative charge is $r = \frac{1}{2}s\sqrt{2}$ and

$$\begin{aligned}
F_E &= 2F \sin 45° = 2\frac{1}{4\pi\varepsilon_0}\frac{q_1 q_2}{r^2}\sin 45°\\[2mm]
&= 2(9 \times 10^9 \text{ N·m}^2/\text{C}^2)\frac{(1.5 \times 10^{-6}\text{ C})(6.2 \times 10^{-9}\text{ C})}{(\frac{1}{2}\cdot 4.0 \times 10^{-2}\cdot \sqrt{2}\text{ m})^2}\sin 45°\\[2mm]
&= 0.15 \text{ N}
\end{aligned}$$

The direction of the net force is straight upward, toward the center of the line that joins the two positive charges.

> **Example 4** Two pith balls of mass m are each given a charge of $+q$. They are hung side-by-side from two threads each of length L, and move apart as a result of their electrical repulsion. Find the equilibrium separation distance x in terms of m, q, and L. (Use the fact that if θ is small, then $\tan \theta \approx \sin \theta$.)

Solution. Three forces act on each ball: weight, tension, and electrical repulsion:

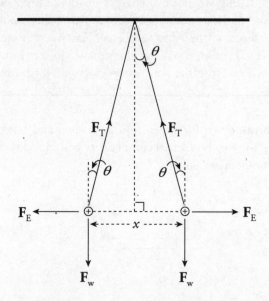

When the balls are in equilibrium, the net force each feels is zero. Therefore, the vertical component of $\mathbf{F}_T$ must cancel out $\mathbf{F}_w$ and the horizontal component of $\mathbf{F}_T$ must cancel out $\mathbf{F}_E$:

$$F_T \cos \theta = F_w \qquad \text{and} \qquad F_T \sin \theta = F_E$$

Dividing the second equation by the first, we get $\tan \theta = F_E/F_w$. Therefore,

$$\tan \theta = \frac{k \dfrac{q^2}{x^2}}{mg} = \frac{kq^2}{mgx^2}$$

Now, to approximate: If θ is small, the $\tan \theta \approx \sin \theta$ and, from the diagram $\sin \theta = \frac{1}{2} x/L$. Therefore, the equation above becomes

$$\frac{\frac{1}{2}x}{L} = \frac{kq^2}{mgx^2} \quad \Rightarrow \quad \frac{1}{2}mgx^3 = kq^2L \quad \Rightarrow \quad x = \sqrt[3]{\frac{2kq^2L}{mg}}$$

THE ELECTRIC FIELD

The presence of a massive body such as the earth causes objects to experience a gravitational force directed toward the earth's center. For objects located outside the earth, this force varies inversely with the square of the distance and directly with the mass of the gravitational source. A vector diagram of the gravitational field surrounding the earth looks like this:

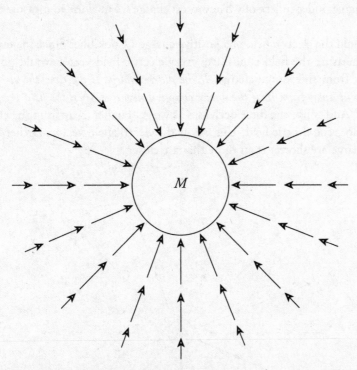

We can think of the space surrounding the earth as permeated by a **gravitational field** created by the earth. Any mass placed in this field then experiences a gravitational force due to an interaction with this field.

The same process is used to describe the electric force. Rather than having two charges reach out across empty space to each other to produce a force, we will instead interpret the interaction in the following way: The presence of a charge creates an **electric field** in the space that surrounds it. Another charge placed in the field created by the first will experience a force due to the field.

Consider a point charge Q in a fixed position and assume that it's positive. Now imagine moving a tiny positive test charge q around to various locations near Q. At each location, measure the force that the test charge experiences, and call it $\mathbf{F}_{on\,q}$. Divide this force by the test charge q; the resulting vector is the **electric field vector**, $\mathbf{E}$, at that location:

$$\mathbf{E} = \frac{\mathbf{F}_{on\,q}}{q}$$

The reason for dividing by the test charge is simple. If we were to use a different test charge with, say, twice the charge of the first one, then each of the forces $\mathbf{F}$ we'd measure would be twice as much as before. But when we divided this new, stronger force by the new, greater test charge, the factors of 2 would cancel, leaving the same ratio as before. So this ratio tells us the intrinsic strength of the field due to the source charge, independent of whatever test charge we may use to measure it.

What would the electric field of a positive charge Q look like? Since the test charge used to measure the field is positive, every electric field vector would point radially away from the source charge. *If the source charge is positive, the electric field vectors point away from it; if the source charge is negative, then the field vectors point toward it.* And, since the force decreases as we get farther away from the charge (as $1/r^2$), so does the electric field. This is why the electric field vectors farther from the source charge are shorter than those that are closer.

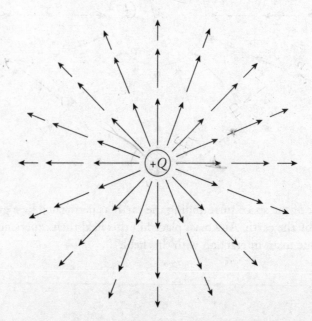

Since the force on any test charge q, due to some source Q, has a strength of $qQ/4\pi\varepsilon_0 r^2$, when we divide this by q, we get the expression for the strength of the electric field created by a point-charge source of magnitude Q:

$$E = \frac{1}{4\pi\varepsilon_0}\frac{Q}{r^2}$$

This is the electric field in the space surrounding for a **point charge**.

To make it easier to sketch an electric field, lines are drawn through the vectors such that the electric field vector is always tangent to the line everywhere it's drawn.

Now, your first thought might be that obliterating the individual field vectors deprives us of information, since the length of the field vectors told us how strong the field was. Well, although the individual field vectors are gone, the strength of the field can be figured out by looking at the density of the field lines. Where the field lines are closer together, the field is stronger.

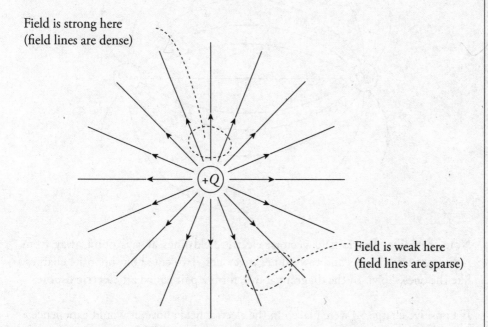

Field is strong here
(field lines are dense)

Field is weak here
(field lines are sparse)

Electric fields obey the same addition properties as the electric force. If we had two source charges, their fields would overlap and effectively add; a third charge wandering by would feel the effect of all three fields. At each position in space, add the electric field vector due to one of the charges to the electric field vector due to the other charge: $\mathbf{E}_{total} = \mathbf{E}_1 + \mathbf{E}_2$. In the diagram below, $\mathbf{E}_1$ is the electric field vector at a particular location due to the charge $+Q$, and $\mathbf{E}_2$ is the electric field vector at that same location due to the other charge, $-Q$. Adding these vectors gives the overall field vector $\mathbf{E}_{\text{total}}$ at that location.

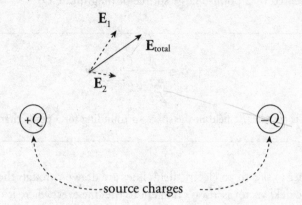

If this is done at enough locations, the electric field lines can be sketched.

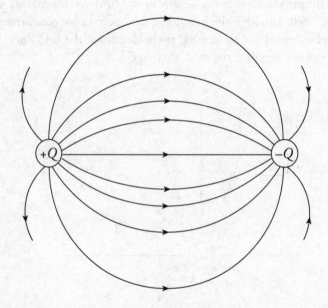

Note that, like electric field vectors, electric field lines always point away from positive source charges and toward negative ones. Two equal but opposite charges, like the ones shown in the diagram above, form a pair called an **electric dipole**.

If a positive charge $+q$ were placed in the electric field above, it would experience a force that is tangent to, and in the same direction as, the field line passing through $+q$'s location. After all, electric fields are sketched from the point of view of what a positive test charge would do. On the other hand, if a negative charge $-q$ were placed in the electric field, it would experience a force that is tangent to, but in the direction opposite from, the field line passing through $-q$'s location.

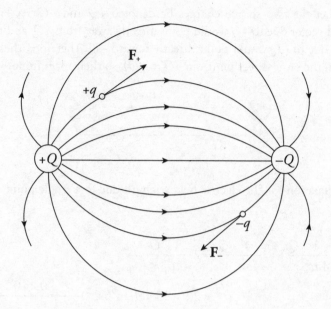

Finally, notice that electric field lines never cross.

Example 5 A charge $q = +3.0$ nC is placed at a location at which the electric field strength is 400 N/C. Find the force felt by the charge q.

Solution. From the definition of the electric field, we have the following equation:

$$\mathbf{F}_{on\ q} = q\mathbf{E}$$

Therefore, in this case, $F_{on\ q} = qE = (3 \times 10^{-9}\ \text{C})(400\ \text{N/C}) = 1.2 \times 10^{-6}\ \text{N}$.

Example 6 A dipole is formed by two point charges, each of magnitude 4.0 nC, separated by a distance of 6.0 cm. What is the strength of the electric field at the point midway between them?

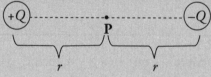

Solution. Let the two source charges be denoted $+Q$ and $-Q$. At Point P, the electric field vector due to $+Q$ would point directly away from $+Q$, and the electric field vector due to $-Q$ would point directly toward $-Q$. Therefore, these two vectors point in the same direction (from $+Q$ to $-Q$), so their magnitudes would add.

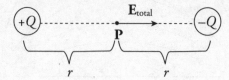

Using the equation for the electric field strength due to a single point charge, we find that

$$E_{total} = \frac{1}{4\pi\varepsilon_0}\frac{Q}{r^2} + \frac{1}{4\pi\varepsilon_0}\frac{Q}{r^2} = 2\frac{1}{4\pi\varepsilon_0}\frac{Q}{r^2}$$

$$= 2(9\times10^9 \text{ N}\cdot\text{m}^2/\text{C}^2)\frac{4.0\times10^{-9}\text{ C}}{\frac{1}{2}(6.0\times10^{-2}\text{ m})^2}$$

$$= 8.0\times10^4 \text{ N/C}$$

Example 7 If a charge $q = -5.0$ pC were placed at the midway point described in the previous example, describe the force it would feel. ("p" is the abbreviation for "pico-," which means 10^{-12}.)

Solution. Since the field E at this location is known, the force felt by q is easy to calculate:

$$\mathbf{F}_{on\,q} = q\mathbf{E} = (-5.0\times10^{-12}\text{ C})(8.0\times10^4\text{ N/C to the right}) = 4.0\times10^{-7}\text{ N to the } \textit{left}$$

Example 8 What can you say about the electric force that a charge would feel if it were placed at a location at which the electric field was zero?

Solution. Remember that $\mathbf{F}_{on\,q} = q\mathbf{E}$. So if E = 0, then $\mathbf{F}_{on\,q} = 0$. (Zero field means zero force.)

The Uniform Electric Field

An important subset of problems deal with uniform electric fields. One method of creating this uniform field is to have two large conducting sheets, each storing some charge Q, some distance d apart. Near the edges of each sheet, the field may not be uniform, but near the middle, for all practical purposes the field is uniform. Having a uniform field means having a constant force (and therefore a constant acceleration), so you can use kinematic equations just as if you had a uniform gravitational field (as you do near the surface of the earth).

Example 9 Positive charge is distributed uniformly over a large, horizontal plate, which then acts as the source of a vertical electric field. An object of mass 5 g is placed at a distance of 2 cm above the plate. If the strength of the electric field at this location is 10^6 N/C, how much charge would the object need to have in order for the electrical repulsion to balance the gravitational pull?

Solution. Clearly, since the plate is positively charged, the object would also have to carry a positive charge so that the electric force would be repulsive.

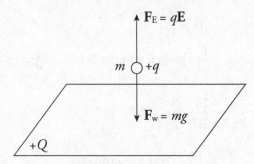

Let q be the charge on the object. Then, in order for F_E to balance mg, we must have

$$qE = mg \quad \Rightarrow \quad q = \frac{mg}{E} = \frac{(5 \times 10^{-3} \text{ kg})(10 \text{ N/kg})}{10^6 \text{ N/C}} = 5 \times 10^{-8} \text{ C} = 50 \text{ nC}$$

Example 10 A proton, neutron, and electron are in a uniform electric field of 20 N/C that is caused by two large charged plates that are 30 cm apart. The particles are far enough apart that they don't interact with each other. They are released from rest equidistant from each plate.

(a) What is the magnitude of the net force acting on each particle?

(b) What is the magnitude of the acceleration of each particle?

(c) How much work will be done on the particle when it collides with one of the charged plates?

(d) What is the speed of each particle when it strikes the plate?

(e) How long does it take to reach the plate?

Solution.

(a) Since $E = \dfrac{F}{q}$, $F = qE$. Plugging in the values, we get

proton: $F = (1.6 \times 10^{-19}\,C)\,(20\,N/C) = 3.2 \times 10^{-18}\,N$

electron: $F = (1.6 \times 10^{-19}\,C)\,(20\,N/C) = 3.2 \times 10^{-18}\,N$

neutron: $F = (0\,C)\,(20\,N/C) = 0\,N$

Note: Because the proton and electron have the same magnitude, they will experience the same force. If you're asked for the direction, the proton travels in the same direction as the electric field and the electron travels in the opposite direction as the electric field.

(b) Since $F = ma$, $a = \dfrac{F}{m}$. Plugging in the values, we get

proton: $a = \dfrac{3.2 \times 10^{-18}\,N}{1.67 \times 10^{-27}\,kg} = 1.9 \times 10^{9}\,m/s^2$

electron: $a = \dfrac{3.2 \times 10^{-18}\,N}{9.11 \times 10^{-31}\,kg} = 3.5 \times 10^{12}\,m/s^2$

neutron: $a = \dfrac{0\,N}{1.67 \times 10^{-27}\,kg} = 0\,m/s^2$

Notice that although the charges have the same magnitude of force, the electron experiences an acceleration almost 2000 times greater due to its mass being almost 2000 times smaller than the proton's mass.

(c) Since $W = Fd$, we get $W = qEd$. Plugging in the values, recalling that the charges start midway between the plates, we get

proton: $W = (1.6 \times 10^{-19}\,C)\,(20\,N/C)\,(0.15\,m) = 4.8 \times 10^{-19}\,N$

electron: $W = (1.6 \times 10^{-19}\,C)\,(20\,N/C)\,(0.15\,m) = 4.8 \times 10^{-19}\,N$

neutron: $W = (0\,C)\,(20\,N/C)\,(0.15\,m) = 0\,N$

(d) Recall one of the big five kinematics equations:

$$v_f^{\,2} = v^{\,2} + 2a(x - x_o) \rightarrow v_f = \sqrt{2a(x - x_o)}$$

If the particles are midway between the 30 cm plates, they will travel 0.15 m.

$$\text{proton: } v_f = \sqrt{2(1.9 \times 10^9 \text{ m/s}^2)(0.15 \text{ m})} \rightarrow v_f = 24,000 \text{ m/s}$$

$$\text{electron: } v_f = \sqrt{2(3.5 \times 10^{12} \text{ m/s}^2)(0.15 \text{ m})} \rightarrow v_f = 1.0 \times 10^6 \text{ m/s}$$

$$\text{neutron: } v_f = \sqrt{2(0 \text{ m/s}^2)(0.15 \text{ m})} \rightarrow v_f = 0 \text{ m/s}$$

Notice that, even though the force is the same and the same work is done on both charges, there is a significant difference in final velocities due to the large mass difference. An alternative solution to this would be using $W = \Delta KE$. You would have obtained the same answers.

(e)   Recall one of the big five kinematics equations:

$$v_f = v_i + at \rightarrow t = \frac{v_f - v_i}{a} \rightarrow t = \frac{v_f}{a}$$

$$\text{proton: } t = \frac{24,000 \text{ m/s}}{1.9 \times 10^9 \text{ m/s}^2} \rightarrow 1.3 \times 10^{-5} \text{ s}$$

$$\text{electron: } t = \frac{1 \times 10^6 \text{ m/s}}{3.5 \times 10^{12} \text{ m/s}^2} \rightarrow 2.9 \times 10^{-7} \text{ s}$$

neutron: The neutron never accelerates, so it will never hit the plate

CONDUCTORS AND INSULATORS

Materials can be classified into broad categories based on their ability to permit the flow of charge. If electrons were placed on a metal sphere, they would quickly spread out and cover the outside of the sphere uniformly. These electrons would be free to flow through the metal and redistribute themselves, moving to get as far away from each other as they could. Materials that permit the flow of excess charge are called **conductors**; they conduct electricity. Metals are the best examples of conductors. Aqueous solutions that contain dissolved electrolytes (such as salt water) are also conductors. Metals conduct electricity because they bind all but their outermost electron very tightly. That outermost electron is free to move about the metal. This creates a sort of sea of mobile (or conduction) electrons.

Insulators, on the other hand, closely guard their electrons—and even extra ones that might be added. Electrons are not free to roam throughout the atomic lattice. Examples of insulators are glass, wood, rubber, and plastic. If excess charge is placed on an insulator, it stays put.

> **Example 11** A solid sphere of copper is given a negative charge. Discuss the electric field inside and outside the sphere.

Solution. The electric field inside a conductor is zero. Therefore, the excess electrons that are deposited on the sphere move quickly to the outer surface (copper is a great conductor). *Any excess charge on a conductor resides entirely on the outer surface.*

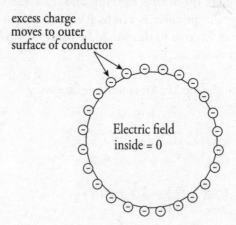

Once these excess electrons establish a uniform distribution on the outer surface of the sphere, there will be no net electric field within the sphere. Why not? Since there is no additional excess charge inside the conductor, there are no excess charges to serve as a source or sink of an electric field line cutting down into the sphere, because field lines begin or end on excess charges. *There can be no electrostatic field within the body of a conductor.*

In fact, you can shield yourself from electric fields simply by surrounding yourself with metal. Charges may move around on the outer surface of your cage, but within the cage, the electric field will be zero. For points outside the sphere, it can be shown that the sphere behaves as if all its excess charge were concentrated at its center. (Remember that this is just like the gravitational field due to a uniform spherical mass.) Also, *the electric field is always perpendicular to the surface, no matter what shape the surface may be.* See the diagram below.

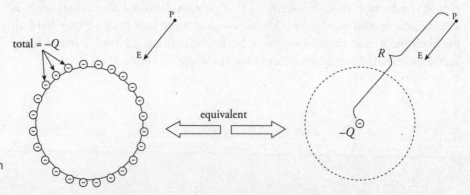

Now let's take our previous sphere and put it in a different situation: Start with a neutral metal sphere and bring a charge (a postive charge) Q nearby without touching the original metal sphere. What will happen? The postive charge will attract free electrons in the metal, leaving the far side of the sphere positively charged. Since the negative charge is closer to Q than the positive charge, there will be a net attraction between Q and the sphere. So, even though the sphere as a whole is electrically neutral, the separation of charge induced by the presence of Q will create a force of electrical attraction between them.

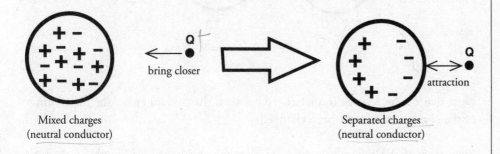

Mixed charges
(neutral conductor)

Separated charges
(neutral conductor)

Charging by Induction

This process may be used to redistribute charges among a pair of neutrally charged spheres, so that in the end both spheres are charged. Imagine two neutrally charged spheres that are each set on an insulating stand. The spheres are arranged so that they are in contact with one another.

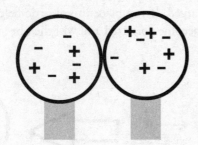

A charge is brought near the side of one of the spheres resulting in the same attraction as occurred with the single sphere. However, because we have two conducting spheres, the excess negative charges are located on the right sphere and the positive excess charges are on the left sphere.

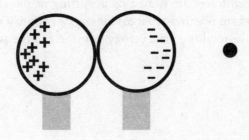

With the external charge still nearby, the two constructing spheres are separated. Because the conductors are no longer in contact, the negative charge has been trapped on the right sphere and the positive charge has been trapped on the left sphere.

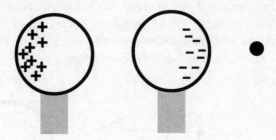

Note that the net charge is unchanged between the two spheres—the distribution of the charges has simply been changed.

Now what if the sphere was made of glass (an insulator)? Although there aren't free elctrons that can move to the near side of the sphere, the atoms that make up the sphere will become polarized. That is, their electrons will feel a tug toward Q, and so will spend more time on the side of the atom closer to Q than on the side opposite Q, causing the atoms to develop a partial negative charge pointing toward Q (and a partial positive charge pointing away from Q). The effect isn't as dramatic as the mass movement of free electrons in the case of a metal sphere, but the polarization is still enough to cause an electrical attraction between the sphere and Q. For example, if you comb your hair, the comb will pick up extra electrons, making it negatively charged. If you place this electric field source near little bits of paper the paper will become polarized and will then be attracted to the comb.

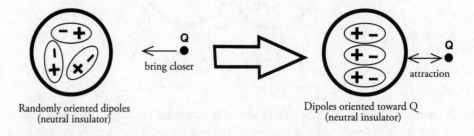

Randomly oriented dipoles (neutral insulator)   bring closer   Dipoles oriented toward Q (neutral insulator)   attraction

The same phenomenon, in which the presence of a charge tends to cause polarization in a nearby collection of charges, is responsible for a kind of intermolecular force. Dipole-induced forces are caused by a shifting of the electron cloud of a neutral molecule toward positively charged ions or away from a negatively charged ion; in either case, the resulting force between the ion and the atom is attractive.

Chapter 5 Review Questions

Solutions can be found in Chapter 12.

Section I: Multiple Choice

1. If the distance between two positive point charges is tripled, then the strength of the electrostatic repulsion between them will decrease by a factor of

   (A) 3
   (B) 6
   (C) 8
   (D) 9

2. Two 1 kg spheres each carry a charge of magnitude 1 C. How does F_E, the strength of the electric force between the spheres, compare to F_G, the strength of their gravitational attraction?

   (A) $F_E < F_G$
   (B) $F_E = F_G$
   (C) $F_E > F_G$
   (D) If the charges on the spheres are of the same sign, then $F_E > F_G$; but if the charges on the spheres are of the opposite sign, then $F_E < F_G$.

3. The figure below shows three point charges, all positive. If the net electric force on the center charge is zero, what is the value of y/x ?

   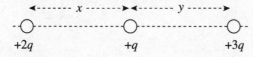

   +2q   +q   +3q

   (A) $\dfrac{4}{9}$

   (B) $\sqrt{\dfrac{2}{3}}$

   (C) $\sqrt{\dfrac{3}{2}}$

   (D) $\dfrac{3}{2}$

4.

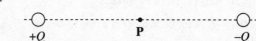

   +Q   P   –Q

   The figure above shows two point charges, +Q and –Q. If the negative charge were absent, the electric field at Point P due to +Q would have strength E. With –Q in place, what is the strength of the total electric field at P, which lies at the midpoint of the line segment joining the charges?

   (A) 0

   (B) $\dfrac{E}{2}$

   (C) E

   (D) 2E

5. A sphere of charge +Q is fixed in position. A smaller sphere of charge +q is placed near the larger sphere and released from rest. The small sphere will move away from the large sphere with

   (A) decreasing velocity and decreasing acceleration
   (B) decreasing velocity and increasing acceleration
   (C) increasing velocity and decreasing acceleration
   (D) increasing velocity and increasing acceleration

6. An object of charge +q feels an electric force $\mathbf{F}_E$ when placed at a particular location in an electric field, **E**. Therefore, if an object of charge –2q were placed at the same location where the first charge was, it would feel an electric force of

   (A) $\dfrac{-\mathbf{F}_E}{2}$

   (B) $-2\mathbf{F}_E$

   (C) $-2q\mathbf{F}_E$

   (D) $\dfrac{-2\mathbf{F}_E}{q}$

7. A charge of $-3Q$ is transferred to a solid metal sphere of radius r. How will this excess charge be distributed?

(A) $-Q$ at the center, and $-2Q$ on the outer surface

(B) $-3Q$ at the center

(C) $-3Q$ on the outer surface

(D) $-Q$ at the center, $-Q$ in a ring of radius $\frac{1}{2}r$, and $-Q$ on the outer surface

Section II: Free Response

1. In the figure shown, all four charges ($+Q$, $+Q$, $-q$, and $-q$) are situated at the corners of a square. The net electric force on each charge $+Q$ is zero.

    (a)   Express the magnitude of q in terms of Q.

    (b)   Is the net electric force on each charge $-q$ also equal to zero? Justify your answer.

    (c)   Determine the electric field at the center of the square.

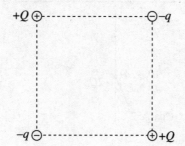

2. Two charges, +Q and +2Q, are fixed in place along the y-axis of an x-y coordinate system as shown in the figure below. Charge 1 is at the point $(0, a)$, and Charge 2 is at the point $(0, -2a)$.

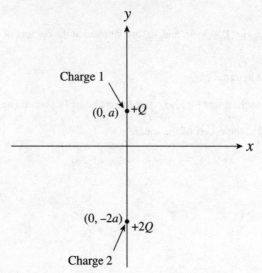

(a) Find the electric force (magnitude and direction) felt by Charge 1 due to Charge 2.

(b) Find the electric field (magnitude and direction) at the origin created by both Charges 1 and 2.

(c) Is there a point on the x-axis where the total electric field is zero? If so, where? If not, explain briefly.

(d) Is there a point on the y-axis where the total electric field is zero? If so, where? If not, explain briefly.

(e) If a small negative charge, $-q$, of mass m were placed at the origin, determine its initial acceleration (magnitude and direction).

Summary

○ Coulomb's Law describes the force acting on two point charges and is given by

$$F_e = \frac{1}{4\pi\varepsilon_0} \frac{q_1 q_2}{r^2}$$

where $\frac{1}{4\pi\varepsilon_0}$ is a constant equal to 9.0×10^9 N·m²/C². You can use all the strategies you used in the Newton's Laws chapter to solve these types of problems.

○ The electric field is given by $E = \frac{F}{q}$

○ The electric field from a point charge is $E = \frac{1}{4\pi\varepsilon_0} \frac{q}{r^2}$

○ Both the electric force and field are vector quantities and therefore all the rules for vector addition apply.

126

Chapter 6
Electronic Potential
and Capacitance

INTRODUCTION

When an object moves in a gravitational field, it usually experiences a change in kinetic energy and in gravitational potential energy due to the work done on the object by gravity. Similarly, when a charge moves in an electric field, it generally experiences a change in kinetic energy and in electrical potential energy due to the work done on it by the electric field. By exploring the idea of electric potential, we can simplify our calculations of work and energy changes within electric fields.

ELECTRICAL POTENTIAL ENERGY

When a charge moves in an electric field, then unless its displacement is always perpendicular to the field, the electric force does work on the charge. If W_E is the work done by the electric force, then the change in the charge's **electrical potential energy** is defined by

$$\Delta U_E = -W_E$$

Notice that this is the same equation that defined the change in the gravitational potential energy of an object of mass m undergoing a displacement in a gravitational field ($\Delta U_G = -W_G$).

Example 1 A positive charge $+q$ moves from position A to position B in a uniform electric field **E**:

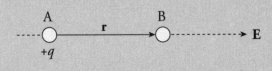

What is its change in electrical potential energy?

Solution. Since the field is uniform, the electric force that the charge feels, $\mathbf{F}_E = q\mathbf{E}$, is constant. Since q is positive, $\mathbf{F}_E$ points in the same direction as **E**, and, as the figure shows, they point in the same direction as the displacement, r. This makes the work done by the electric field equal to $W_E = F_E r = qEr$, so the change in the electrical potential energy is

$$\Delta U_E = -qEr$$

Note that the change in potential energy is negative, which means that potential energy has decreased; this always happens when the field does positive work. It's just like dropping a rock to the ground: Gravity does positive work, and the rock loses gravitational potential energy.

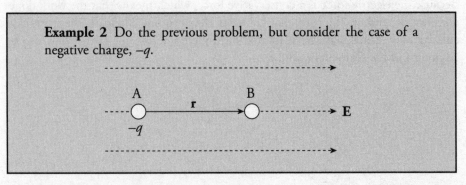

Example 2 Do the previous problem, but consider the case of a negative charge, $-q$.

Solution. In this case, an outside agent must be pushing the charge to make it move, because the electric force *naturally* pushes negative charges against field lines. Therefore, we expect that the work done by the electric field is negative. The electric force, $\mathbf{F}_E = (-q)\mathbf{E}$, points in the direction opposite to the displacement, so the work it does is $W_E = -F_E r = -qEr = -qEr$. Thus, the change in electrical potential energy is positive: $\Delta U_E = -W_E = -(-qEr) = qEr$. Because the change in potential energy is positive, the potential energy increases; this always happens when the field does negative work. It's like lifting a rock off the ground: Gravity does negative work, and the rock gains gravitational potential energy.

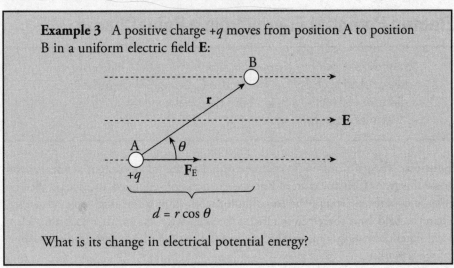

Example 3 A positive charge $+q$ moves from position A to position B in a uniform electric field $\mathbf{E}$:

What is its change in electrical potential energy?

Solution. The electric force felt by the charge q is $\mathbf{F}_E = q\mathbf{E}$ and this force is parallel to $\mathbf{E}$, because q is positive. In this case, because $\mathbf{F}_E$ is not parallel to r (as it was in Example 1), we will use the more general definition of work for a constant force:

$$W_E = \mathbf{F}_E \cdot \mathbf{r} = F_E\, r \cos\theta = qEr \cos\theta$$

But $r \cos \theta = d$, so

$$W_E = qEd \text{ and } \Delta U_E = -W_E = -qEd$$

Because the electric force is a conservative force, which means that the work done does not depend on the path that connects the positions A and B, the work calculated above could have been figured out by considering the path from A to B composed of the segments $\mathbf{r}_1$ and $\mathbf{r}_2$:

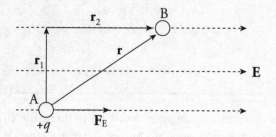

Along $\mathbf{r}_1$, the electric force does no work since this displacement is perpendicular to the force. Thus, the work done by the electric field as q moves from A to B is simply equal to the work it does along $\mathbf{r}_2$. And since the length of $\mathbf{r}_2$ is $d = r \cos \theta$, we have $W_E = F_E d = qEd$, just as before.

Electric Potential Energy from a Point Charge

Example 4 A positive charge $q_1 = +2 \times 10^{-6}$ C is held stationary, while a negative charge, $q_2 = -1 \times 10^{-8}$ C, is released from rest at a distance of 10 cm from q_1. Find the kinetic energy of charge q_2 when it's 1 cm from q_1.

Solution. The gain in kinetic energy is equal to the loss in potential energy; you know this from Conservation of Energy. Previously, we looked at constant electric fields and were able to use the equation for work from a constant force. However, when the field (and therefore the force) changes, we need another equation. Electrical potential energy required to move along the field lines surrounding a point charge is given by

$$U_E = \frac{1}{4\pi\varepsilon_0}\left(\frac{q_1 q_2}{r}\right)$$

Therefore, if q_1 is fixed and q_2 moves from r_A to r_B, the change in potential energy is

$$\Delta U_E = U_B - U_A$$

$$= \frac{q_1}{4\pi\varepsilon_0}\left(\frac{q_2}{r_B} - \frac{q_2}{r_A}\right)$$

$$= \frac{q_1 q_2}{4\pi\varepsilon_0}\left(\frac{1}{r_B} - \frac{1}{r_A}\right)$$

$$= (9 \times 10^9 \text{ N} \cdot \text{m}^2 / \text{C}^2)(+2 \times 10^{-6} \text{ C})(-1 \times 10^{-8} \text{ C})\left(\frac{1}{0.01 \text{ m}} - \frac{1}{0.10 \text{ m}}\right)$$

$$= -0.016 \text{ J}$$

Since q_2 lost 0.016 J of potential energy, the gain in kinetic energy is 0.016 J. Since q_2 started from rest (with no kinetic energy), this is the kinetic energy of q_2 when it's 1 cm from q_1.

Example 5 Two positive charges, q_1 and q_2, are held in the positions shown below. How much work would be required to bring (from infinity) a third positive charge, q_3, and place it so that the three charges form the corners of an equilateral triangle of side length s?

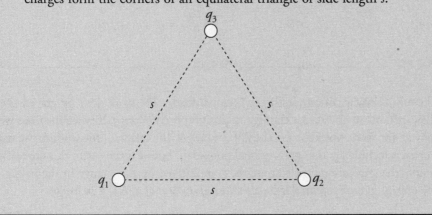

Solution. An external agent would need to do positive work, equal in magnitude to the negative work done by the electric force on q_3 as it is brought into place, so let's first compute this quantity. Let's first compute the work done *by the electric force* as q_3 is brought in. Since q_3 is fighting against both q_1's and q_2's electric field, the total work done on q_3 by the electric force, W_E, is equal to the work done on q_3 by q_1 (W_{1-3}) plus the work done on q_3 by q_2 (W_{2-3}). Using the equation $W_E = -\Delta U_E$ and the one we gave above for ΔU_E, we have

$$W_{1-3} + W_{2-3} = -\ U_{1-3} + -\ U_{2-3}$$

$$= -\frac{q_1 q_3}{4\pi\varepsilon_0}\frac{1}{s} - 0\ +-\frac{q_2 q_3}{4\pi\varepsilon_0}\frac{1}{s} - 0$$

$$= -\frac{1}{4\pi\varepsilon_0}\frac{q_1 q_3}{s} + \frac{1}{4\pi\varepsilon_0}\frac{q_2 q_3}{s}$$

Therefore, the work that an external agent must do to bring q_3 into position is

$$-W_E = \frac{1}{4\pi\varepsilon_0}\frac{q_1 q_3}{s} + \frac{1}{4\pi\varepsilon_0}\frac{q_2 q_3}{s}$$

NOT the Same

Electric Potential and Electric Potential Energy, although very closely related and having simliar names, are not the same thing. Even the units are measured differently—electric potential is measured in joules per coulomb and electric potential energy in joules.

ELECTRIC POTENTIAL

Let W_E be the work done by the electric field on a charge q as it undergoes a displacement. If another charge, say $2q$, were to undergo the same displacement, the electric force would be twice as great on this second charge, and the work done by the electric field would be twice as much, $2W_E$. Since the work would be twice as much in the second case, the change in electrical potential energy would be twice as great as well, but the ratio of the change in potential energy to the charge would be the same: $U_E/q = U_E/2q$. This ratio says something about the *field* and the *displacement* but not the charge that made the move. The change in **electric potential** ΔV, is defined as this ratio:

Equation Sheet

$$V = \frac{U_E}{q}$$

There is a similar concept with gravitational fields. The work done by gravity (near the earth) on an object as it changes its position is $\Delta W = mgh$. If twice the mass were moved, the work would have to be $\Delta W = (2m)gh$. The ration of the work to the mass moved is technically the "gravitational potential," but near the earth, this quantity is simply the gravitational field strength, g, multiplied by the change in the height of the object. *Electric potential is the electrical equivalent of a change in height.*

Electric potential is electrical potential energy *per unit charge*; the units of electric potential are joules per coulomb. One joule per coulomb is called one **volt** (abbreviated V); so 1 J/C = 1 V.

Potential Difference

Finding the potential at a certain single point in space is meaningless. With potential, what is more important is an initial position and final position. Just like in gravity, finding the potential energy at a certain point is meaningless if there is no relation to another point. Once we establish this potential difference, when we move a charge across it we can generate energy. This will be key in the next chapter where potential difference is needed in order to have an electrical circuit.

Electric Potential from a Point Charge

Consider the electric field created by a point source charge Q. If a charge q moves from a distance r_A to a distance r_B from Q, then the change in the potential energy is

$$U_B - U_A = \frac{Qq}{4\pi\varepsilon_0}\left[\frac{1}{r_B} - \frac{1}{r_A}\right]$$

The difference in electric potential between positions A and B in the field created by Q is

$$V_B - V_A = \frac{U_B - U_A}{q} = \frac{Q}{4\pi\varepsilon_0}\left(\frac{1}{r_B} - \frac{1}{r_A}\right)$$

If we designate $V_A \to 0$ as $r_A \to \infty$ (an assumption that's stated on the AP Physics Exam), then the electric potential at a distance r from Q is

$$V = \frac{1}{4\pi\varepsilon_0}\frac{Q}{r}$$

Equation Sheet

Note that the potential depends on the strength of the source charge making the field and the distance from the source charge.

Example 6 Let $Q = 2 \times 10^{-9}$ C. What is the potential at a Point P that is 2 cm from Q?

Solution. Relative to $V = 0$ at infinity, we have

$$V = \frac{1}{4\pi\varepsilon_0}\frac{Q}{r} = \left(9 \times 10^9 \text{ N} \cdot \text{m}^2/\text{C}^2\right)\frac{2 \times 10^{-9} \text{ C}}{0.02 \text{ m}} = 900 \text{ V}$$

This means that the work done by the electric field on a charge of q coulombs brought to a point 2 cm from Q would be $-900q$ joules.

Note that, like potential energy, electric potential is a *scalar*. In the preceding example, we didn't have to specify the direction of the vector from the position of Q to the Point P, because it didn't matter. Imagine a sphere with a surface of 2 cm from Q; at any point on that sphere, the potential will be 900 V. These spheres around Q are called **equipotential surfaces**, and they're surfaces of constant potential. The equipotentials are always perpendicular to the electric field lines.

Remember
Equipotential surfaces are often imaginary. Whether we put a metal sphere in that location or whether we visualize a sphere that isn't really there doesn't change the electric potential in that region.

Example 7 How much work is done by the electric field as a charge moves along an equipotential surface?

Solution. If the charge always remains on a single equipotential, then, by definition, the potential, V, never changes. Therefore, $\Delta V = 0$, so $\Delta U_E = 0$. Since $W_E = -\Delta U_E$, the work done by the electric field is zero.

Addition of Electric Potential

The formula $V = kQ/r$ tells us how to find the potential due to a single point source charge, Q. Potential is scalar (we will not be concerned with orientation just the sign of charge). When we add up individual potentials, we're simply adding numbers; we're not adding vectors.

Just like Gravitational Potential Energy
The electric force is conservative. All we care about is the change in position when calculating the change in potential energy.

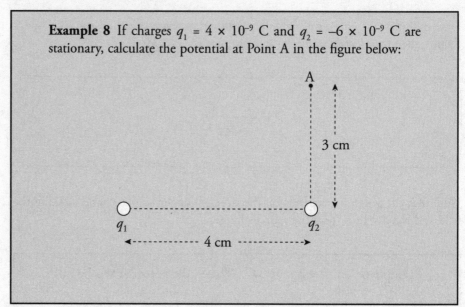

Example 8 If charges $q_1 = 4 \times 10^{-9}$ C and $q_2 = -6 \times 10^{-9}$ C are stationary, calculate the potential at Point A in the figure below:

A

3 cm

q_1 q_2

4 cm

Solution. Potentials add like ordinary numbers. Therefore, the potential at A is just the sum of the potentials at A due to q_1 and q_2. Note that the distance from q_1 to A is 5 cm.

$$V = \frac{1}{4\pi\varepsilon_0} \frac{q_1}{r_{1A}} + \frac{1}{4\pi\varepsilon_0} \frac{q_2}{r_{2A}}$$

$$= \frac{1}{4\pi\varepsilon_0}\left(\frac{q_1}{r_{1A}} + \frac{q_2}{r_{2A}}\right)$$

$$= (9 \times 10^9 \ \text{N}\cdot\text{m}^2/\text{C}^2)\left(\frac{4\times10^{-9}\ \text{C}}{0.05\ \text{m}} + \frac{-6\times10^{-9}\ \text{C}}{0.03\ \text{m}}\right)$$

$$= -1080\ \text{V}$$

Example 9 How much work would it take to move a charge $q = +1 \times 10^{-2}$ C from Point A to Point B (the point midway between q_1 and q_2)? Note: We are using the data from Example 8.

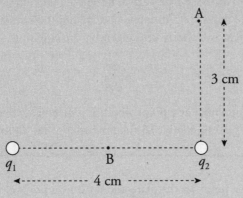

Solution. $\Delta U_E = q\Delta V$, so if we calculate the potential difference between Points A and B and multiply by q, we will have found the change in the electrical potential energy: $\Delta U_{A \to B} = q\Delta V_{A \to B}$. Then, since the work by the electric field is $-\Delta U$, the work required by an external agent is ΔU. In this case, the potential at Point B is

$$V_B = \frac{1}{4\pi\varepsilon_0} \; \frac{q_1}{r_{1B}} + \frac{q_2}{r_{2B}}$$

$$= (9 \times 10^9 \text{ N} \cdot \text{m}^2/\text{C}^2) \; \frac{4 \times 10^{-9} \text{ C}}{0.02 \text{ m}} + \frac{-6 \times 10^{-9} \text{ C}}{0.02 \text{ m}}$$

$$= -900 \text{ V}$$

In the preceding example, we calculated the potential at Point A: $V_A = -1080$ V, so $\Delta V_{A \to B} = V_B - V_A = (-900 \text{ V}) - (-1080 \text{ V}) = +180$ V. This means that the change in electrical potential energy as q moves from A to B is

$$\Delta U_{A \to B} = q\Delta V_{A \to B} \; = (+1 \times 10^{-2} \text{ C})(+180 \text{ V}) = 1.8 \text{ J}$$

This is the work required by an external agent to move q from A to B.

THE ELECTRIC POTENTIAL OF A UNIFORM FIELD

Example 10 Consider a very large, flat plate that contains a uniform surface charge density σ. At points that are not too far from the plate, the electric field is uniform and given by the equation

$$E = \frac{\sigma}{2\varepsilon_0}$$

What is the potential at a point which is a distance d from the sheet (close to the plates), relative to the potential of the sheet itself?

Solution. Let A be a point on the plate and let B be a point a distance d from the sheet. Then

$$V_B - V_A = \frac{-W_{E,\,A \to B} \text{ on } q}{q}$$

Since the field is constant, the force that a charge q would feel is also constant, and is equal to

$$F_E = qE = q\frac{\sigma}{2\varepsilon_0}$$

Therefore,

$$W_{E,\,A \to B} = F_E d$$
$$= \frac{q\sigma}{2\varepsilon_0}d$$

so applying the definition gives us

$$V_B - V_A = \frac{-W_{E,\,A \to B}}{q} = -\frac{\sigma}{2\varepsilon_0}d$$

This says that the potential decreases linearly as we move away from the plate.

Example 11 Two large flat plates—one carrying a charge of $+Q$, the other $-Q$—are separated by a distance d. The electric field between the plates, **E**, is uniform. Determine the potential difference between the plates.

Solution. Imagine a positive charge q moving from the positive plate to the negative plate:

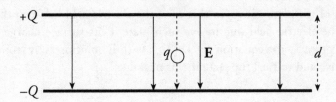

Since the work done by the electric field is

$$W_{E, + \to -} = F_E d = qEd$$

the potential difference between the plates is

$$V_- - V_+ = \frac{-W_{E, + \to -}}{q} = \frac{-qEd}{q} = -Ed$$

This tells us that the potential of the positive plate is greater than the potential of the negative plate, by the amount Ed. This equation can also be written as

$$E = -\frac{V_- - V_+}{d}$$

Therefore, if the potential difference and the distance between the plates are known, then the magnitude of the electric field can be determined quickly. The magnitude is simply

$$E = \left| \frac{\Delta V}{d} \right|$$

Equation Sheet

CAPACITANCE

Consider two conductors, separated by some distance, that carry equal but opposite charges, $+Q$ and $-Q$. Such a pair of conductors comprise a system called a **capacitor**. Work must be done to create this separation of charge, and, as a result, potential energy is stored. Capacitors are basically storage devices for electrical potential energy.

The conductors may have any shape, but the most common conductors are parallel metal plates or sheets. These types of capacitors are called **parallel-plate capacitors**. We'll assume that the distance d between the plates is small compared to the dimensions of the plates since, in this case, the electric field between the plates is uniform. The electric field due to *one* such plate, if its surface charge density is $\sigma = Q/A$, is given by the equation $E = \sigma/(2\varepsilon_0)$, with **E** pointing away from the sheet if σ is positive and toward the plate if σ is negative.

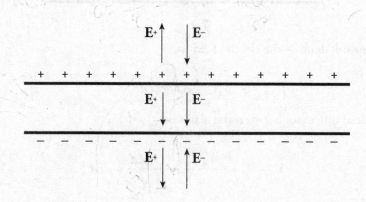

Therefore, with two plates, one with surface charge density $+\sigma$ and the other $-\sigma$, the electric fields combine to give a field that's zero outside the plates and that has the magnitude

$$E_{\text{total}} = \frac{\sigma}{2\varepsilon_0} + \frac{\sigma}{2\varepsilon_0} = \frac{\sigma}{\varepsilon_0}$$

in between.

Equation Sheet

$$E_{total} = \frac{Q}{\varepsilon_0 A}$$

In Example 11, we learned that the magnitude of the potential difference, ΔV, between the plates satisfies the relationship $\Delta V = Ed$, so combining this with the previous equation, we get

$$E = \frac{\sigma}{\varepsilon_0} \implies \frac{\Delta V}{d} = \frac{\sigma}{\varepsilon_0} \implies \frac{\Delta V}{d} = \frac{Q/A}{\varepsilon_0} \implies \frac{Q}{\Delta V} = \frac{\varepsilon_0 A}{d}$$

The River: Dam
In terms of a circuit (which we will cover in the next chapter) we can think of the Capacitor as a dam in a river. Water builds up in one section of the dam before it can overflow and pass to the next section.

What determines capacitance?
Capacitance does NOT determine the charge on the capacitor, Q, or the potential difference, V, across it. It only shows a relationship between Q and V. The greater the voltage applied to a capacitor the greater the amount of charge we can hold on the capacitor. Think of capacitance as a property of a physical object. Physically, capacitance is determined by two things: Area of the plates, A, and the separation between the plates, d.

The ratio of Q to ΔV, for *any* capacitor, is called its **capacitance** (C),

$$C = Q/\Delta V$$

so for a parallel-plate capacitor, we get

$$C = \frac{\varepsilon_0 A}{d}$$

The capacitance measures the capacity for holding charge. The greater the capacitance, the more charge can be stored on the plates at a given potential difference. The capacitance of any capacitor depends only on the size, shape, and separation of the conductors. From the definition, $C = Q/\Delta V$, the units of C are coulombs per volt. One coulomb per volt is renamed one **farad** (abbreviated F): $1 \text{ C/V} = 1 \text{ F}$.

Example 12 A 10-nanofarad parallel-plate capacitor holds a charge of magnitude 50 μC on each plate.

    (a)   What is the potential difference between the plates?

    (b)   If the plates are separated by a distance of 0.2 mm, what is the area of each plate?

Solution.

(a)   From the definition, $C = Q/\Delta V$, we find that

$$V = \frac{Q}{C} = \frac{50 \times 10^{-6} \text{ C}}{10 \times 10^{-9} \text{ F}} = 5000 \text{ V}$$

(b)   From the equation $C = \varepsilon_0 A/d$, we can calculate the area, A, of each plate:

$$A = \frac{Cd}{\varepsilon_0} = \frac{(10 \times 10^{-9} \text{ F})(0.2 \times 10^{-3} \text{ m})}{8.85 \times 10^{-12} \text{ C}^2/\text{N} \cdot \text{m}^2} = 0.23 \text{ m}^2$$

ELECTRIC FIELD AND CAPACITORS

When we studied electric fields in the preceding chapter, we noticed that the electric field created by one or more point source charges varied, depending on the location. For example, as we move further away from the source charge, the electric field gets weaker. Even if we stay at the same distance from, say, a single source charge, the direction of the field changes as we move around. Therefore, we could never obtain an electric field that was constant in both magnitude and direction throughout some region of space from point-source charges. However, the electric field that is created between the plates of a charged parallel-plate capacitor is constant in both magnitude and direction throughout the region between the plates; in other words, a charged parallel-plate capacitor can create a uniform electric field. The electric field, $\mathbf{E}$, always points from the positive plate toward the negative plate, and its magnitude remains the same at every point between the plates, whether we choose a point close to the positive plate, closer to the negative plate, or between them.

Because $\mathbf{E}$ is so straightforward (it's the same everywhere between the plates), the equation for calculating it is equally straightforward, $V = Ed$. The strength of $\mathbf{E}$ depends upon the voltage between the plates, ΔV.

The equation $F = qE$ showed us that the units of E are N/C. The formula $\Delta V = Ed$ tells us that the units of $\mathbf{E}$ are also *V/m*. You'll see both newtons-per-coulomb and volts-per-meter used as units for the electric field; it turns out these units are exactly the same.

> **Example 13** The charge on a parallel-plate capacitor is 4×10^{-6} C. If the distance between the plates is 2 mm and the capacitance is 1 μF, what's the strength of the electric field between the plates?

Solution. Since $C = Q/\Delta V$, we have $\Delta V = Q/C = (4 \times 10^{-6}$ C$)/(10^{-6}$ F$) = 4$ V. Now, using the equation $\Delta V = Ed$,

$$E = \Delta V/d = 4 \ \Delta V/(2 \times 10^{-3} \text{ m}) = 2000 \text{ V/m}$$

> **Example 14** The plates of a parallel-plate capacitor are separated by a distance of 2 mm. The device's capacitance is 1 μF. How much charge needs to be transferred from one plate to the other in order to create a uniform electric field whose strength is 10^4 V/m?

Solution. Because $Q = C\Delta V$ and $\Delta V = Ed$, we find that

$$Q = CEd = (1 \times 10^{-6} \text{ F})(1 \times 10^4 \text{ V/m})(2 \times 10^{-3} \text{ m}) = 2 \times 10^{-5} \text{ C} = 20 \text{ μC}$$

Example 15 A proton (whose mass is m) is placed on top of the positively charged plate of a parallel-plate capacitor, as shown below.

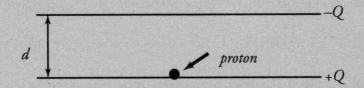

The charge on the capacitor is Q, and the capacitance is C. If the electric field in the region between the plates has magnitude E, give an expression that shows the time required for the proton to move up to the other plate.

Solution. Once we find the acceleration of the proton, we can use Big Five #3, with $v_0 = 0$, to find the time it will take for the proton to move the distance $y = d$. The acceleration of the proton is F/m, where $F = qE$ is the force the proton feels; this gives $a = qE/m$. (We're ignoring the gravitational force on the proton because it is so much weaker than the electric force.) Now, since $E = \Delta V/d$ and $\Delta V = Q/C$, the expression becomes $a = eQ/mdC$ for a. Note that $q = +e$ for a proton. If we consider $y = d$, then Big Five #3 gives us

$$y = v_0 + \frac{1}{2}at^2$$

$$y = \frac{1}{2}at^2$$

$$y = \frac{1}{2}\frac{eQ}{mdC}t^2$$

Solving for t,

$$t = \sqrt{\frac{2dmC}{eQ}}$$

THE ENERGY STORED IN A CAPACITOR

To figure out the electrical potential energy stored in a capacitor, imagine taking a small amount of negative charge off the positive plate and transferring it to the negative plate. This requires that positive work is done by an external agent, and this is the reason that the capacitor stores energy. If the final charge on the capacitor is Q, then we transferred an amount of charge equal to Q, fighting against the prevailing voltage at each stage. If the final voltage is ΔV, then the average voltage during the charging process is $\frac{1}{2}\Delta V$; so, because ΔU_E is equal to charge times voltage, we can write $\Delta U_E = Q \cdot \frac{1}{2}\Delta V = \frac{1}{2}Q\Delta V$. At the beginning of the charging process, when there was no charge on the capacitor, we had $U_i = 0$, so $\Delta U_E = U_f - U_i = U_f - 0 = U_f$; therefore, we have

Equation Sheet

$$U_E = \frac{1}{2}Q\Delta V$$

This is the electrical potential energy stored in a capacitor. Because of the definition $C = Q/\Delta V$, the equation for the stored potential energy can be written as:

$$U_E = \frac{1}{2}(C\Delta V) \cdot \Delta V = \frac{1}{2}C(\Delta V)^2$$

or

$$U_E = \frac{1}{2}Q \cdot \frac{Q}{C} = \frac{Q^2}{2C}$$

CAPACITORS AND DIELETRICS

One method of keeping the plates of a capacitor apart, which is necessary to maintain charge separation and store potential energy, is to insert an insulator (called a **dielectric**) between the plates.

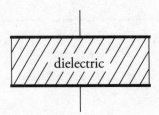

Insulator?
Why not a conductor? If we place a conductor it defeats the purpose of a capacitor because now electrons are free to pass.

> *A dielectric always increases the capacitance of a capacitor.*

Let's see why this is true. Imagine charging a capacitor to a potential difference of ΔV with charge $+Q$ on one plate and $-Q$ on the other. Now disconnect the capacitor from the charging source and insert a dielectric. What happens? Although the dielectric is not a conductor, the electric field that existed between the plates causes the molecules within the dielectric material to polarize; there is more electron density on the side of the molecule near the positive plate.

The effect of this is to form a layer of negative charge along the top surface of the dielectric and a layer of positive charge along the bottom surface; this separation of charge induces its own electric field (E_i), within the dielectric, which opposes the original electric field, E, within the capacitor.

So the overall electric field has been reduced from its previous value: $E_{total} = E + E_i$, and $E_{total} = E - E_i$. Let's say that the electric field has been reduced by a factor of κ (the Greek letter *kappa*) from its original value as follows

$$E_{\text{with dielectric}} = E_{\text{without dielectric}} - E_i = \frac{E}{\kappa}$$

Since $\Delta V = Ed$ for a parallel-plate capacitor, we see that ΔV must have decreased by a factor of κ. But $C = \dfrac{Q}{\Delta V}$, so if ΔV decreases by a factor of κ, then C increases by a factor of κ:

$$C_{\text{with dielectric}} = \kappa C_{\text{without dielectric}}$$

The value of κ, called the **dielectric constant**, varies from material to material, but it's always greater than 1. In general, the capacitance of parallel-plate capacitors is:

$$C = \kappa \varepsilon_0 \, A/d$$

Chapter 6 Review Questions

Solutions can be found in Chapter 12.

Section I: Multiple Choice

1. If the electric field does negative work on a negative charge as the charge undergoes a displacement from Position A to Position B within an electric field, then the electrical potential energy

   (A) is negative
   (B) is positive
   (C) increased
   (D) decreased

2.

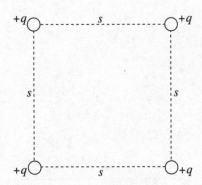

   The work required to assemble the system shown above, bringing each charge in from an infinite distance, is equal to

   (A) $\dfrac{1}{4\pi\varepsilon_0}\dfrac{4q^2}{s}$

   (B) $\dfrac{1}{4\pi\varepsilon_0}\dfrac{(4+\sqrt{2})q^2}{s}$

   (C) $\dfrac{1}{4\pi\varepsilon_0}\dfrac{6q^2}{s}$

   (D) $\dfrac{1}{4\pi\varepsilon_0}\dfrac{(4+2\sqrt{2})q^2}{s}$

3. Negative charges are accelerated by electric fields toward points

   (A) at lower electric potential
   (B) at higher electric potential
   (C) where the electric field is weaker
   (D) where the electric field is stronger

4. A charge q experiences a displacement within an electric field from Position A to Position B. The change in the electrical potential energy is ΔU_E, and the work done by the electric field during this displacement is W_E. Then

   (A) $V_A - V_B = qW_E$
   (B) $V_B - V_A = qW_E$
   (C) $V_A - V_B = \Delta U_E/q$
   (D) $V_B - V_A = \Delta U_E/q$

5.

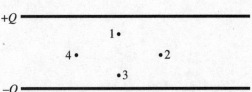

   Which points in this uniform electric field (between the plates of the capacitor) shown above lie on the same equipotential?

   (A) 1 and 3 only
   (B) 2 and 4 only
   (C) 3 and 4 only
   (D) 1, 2, 3, and 4 all lie on the same equipotential since the electric field is uniform.

6.

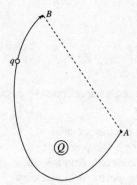

How much work would the electric field (created by the stationary charge Q) perform as a charge q is moved from Point A to B along the curved path shown? $V_A = 200$ V, $V_B = 100$ V, $q = -0.05$ C, length of line segment $AB = 10$ cm, length of curved path $= 20$ cm.

(A) -10 J
(B) -5 J
(C) $+5$ J
(D) $+10$ J

Section II: Free Response

1. In the figure shown, all four charges are situated at the corners of a square with sides s.

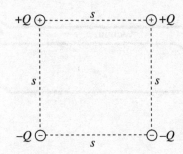

(a) What is the total electrical potential energy of this array of fixed charges?

(b) What is the electric field at the center of the square?

(c) What is the electric potential at the center of the square?

(d) Sketch (on the diagram) the portion of the equipotential surface that lies in the plane of the figure and passes through the center of the square.

(e) How much work would the electric field perform on a charge q as it moved from the midpoint of the right side of the square to the midpoint of the top of the square?

2. The figure below shows a parallel-plate capacitor. Each rectangular plate has length L and width w, and the plates are separated by a distance d.

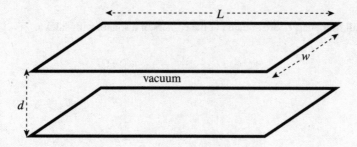

(a)   Determine the capacitance.

An electron (mass m, charge $-e$) is shot horizontally into the empty space between the plates, midway between them, with an initial velocity of magnitude v_0. The electron just barely misses hitting the end of the top plate as it exits. (Ignore gravity.)

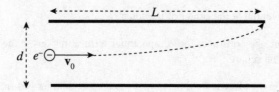

(b)   In the diagram, sketch the electric field vector at the position of the electron when it has traveled a horizontal distance of $L/2$.

(c)   In the diagram, sketch the electric force vector on the electron at the same position as in part (b).

(d)   Determine the strength of the electric field between the plates. Write your answer in terms of L, d, m, e, and v_0.

(e)   Determine the charge on the top plate.

(f)   How much potential energy is stored in the capacitor?

3. A solid conducting sphere of radius a carries an excess charge of Q.

   (a) Determine the electric field magnitude, $E(r)$, as a function of r, the distance from the sphere's center.

   (b) Determine the potential, $V(r)$, as a function of r. Take the zero of potential at $r = \infty$.

   (c) On the diagrams below, sketch $E(r)$ and $V(r)$. (Cover at least the range $0 < r < 2a$.)

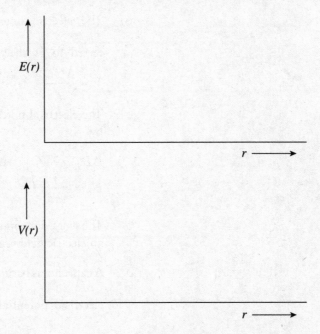

Summary

- The electric potential (commonly referred to as the voltage) is defined as $V = \dfrac{\Delta U_E}{q}$.

- The electrical potential energy is defined by

$$\Delta U_E = -W_E \quad \text{or} \quad U_E = qV \quad \text{or}$$
$$U_E = \frac{1}{4\pi\varepsilon_0}\frac{q_1 q_2}{r}.$$

- The work done moving a charge through an electric field is given by $W = qEd$.

- A capacitor is basically a storage device for electrical potential energy. Capacitance is given by $C = \dfrac{Q}{V}$. For parallel plate capacitors, we have $C = \dfrac{\varepsilon_0 A}{d}$ where $\varepsilon_0 = 8.85 \times 10^{-12}$ C²/Nm², A is the area of the plate and d is the distance between the plates.

- The electrical energy stored in a capacitor is given by either

$$U_C = \frac{1}{2}QV \text{ or } U_C = \frac{1}{2}CV^2.$$

Chapter 7
Direct Current
Circuits

INTRODUCTION

Today, electricity is well-known and appears all around us, from the lighting in our houses to the computing in our personal computers. All of these are powered by complex circuits. In this chapter, we will study the basics of simple direct current circuits.

ELECTRIC CURRENT

A Note on Convention for this Chapter
In this chapter, we will use the standard convention of referring to a change in potential as a voltage, V. In the previous chapter, we referred to these changes in potential as ΔV. Make note of this as you move forward so as to not be confused.

Picture a piece of metal wire. Within the metal, electrons are zooming around at speeds of about a million m/s in random directions, colliding with other electrons and positive ions in the lattice. This constitutes charge in motion, but it doesn't constitute *net* movement of charge, because the electrons move randomly. If there's no net motion of charge, there's no current. However, if we were to create a potential difference between the ends of the wire, meaning if we set up an electric field, the electrons would experience an electric force, and they would start to drift through the wire. This is current. Although the electric field would travel through the wire at nearly the speed of light, the electrons themselves would still have to make their way through a crowd of atoms and other free electrons, so their **drift speed**, v_d, would be relatively slow, on the order of a millimeter per second.

The River Current
Think of the current as the river. In the river we are moving water at a rate; in a current we are moving electric charge at a rate.

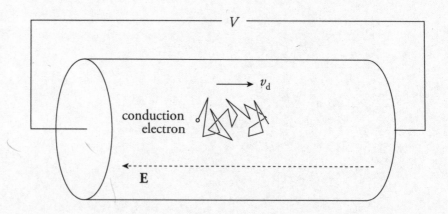

To measure the current, we have to measure how much charge crosses a plane per unit time. If an amount of charge of magnitude ΔQ crosses an imaginary plane in a time interval Δt, then the **average current** is

Equation Sheet

$$I_{\text{avg}} = \frac{\Delta Q}{\Delta t}$$

Because current is charge per unit time, it's expressed in coulombs per second. One coulomb per second is an **ampere** (abbreviated **A**), or amp. So 1 C/s = 1 A.

Although the charge carriers that constitute the current within a metal are electrons, the direction of the current is taken to be the direction that *positive* charge carriers would move. (This is explicitly stated on the AP Physics Exam.) So, if the conduction electrons drift to the right, we'd say the current points toward the left.

RESISTANCE

Let's say we had a copper wire and a glass fiber that had the same length and cross-sectional area, and that we hooked up the ends of the metal wire to a source of potential difference and measured the resulting current. If we were to do the same thing with the glass fiber, the current would probably be too small to measure, but why? Well, the glass provided more resistance to the flow of charge. If the potential difference is V and the current is I, then the **resistance** is

$$R = \frac{V}{I} \text{ or } V = IR$$

Equation Sheet

This is known as Ohm's Law. Not all devices are ohmic, but many are. Notice for the same voltage if the current is large, the resistance is low, and if the current is small, then resistance is high. Because resistance is voltage divided by current, it is expressed in volts per amp. One volt per amp is one **ohm** (Ω, *omega*). So, 1 V/A = 1 Ω.

RESISTIVITY

The resistance of an object depends on two things: the material it's made of and its shape. For example, think again of the copper wire and glass fiber of the same length and area. They have the same shape, but their resistances are different because they're made of different materials. Glass has a much greater intrinsic resistance than copper does; it has a greater **resistivity**. Each material has its own characteristic resistivity, and resistance depends on how the material is shaped. For a wire of length L and cross-sectional area A made of a material with resistivity ρ, resistance is given by:

Benjamin Franklin
Conventional direction of current is "wrong" because Benjamin Franklin made an error when he first started investigating electrostatics. Current direction is in the flow of positive charges, but electrons flow opposite to this convention.

Resistance vs Resistivity
Resistance and Resistivity are NOT the same. Resistance is the extent to which a resistor impedes the passage of electric current. Resistivity is a measure of a material's intrinsic resistance. The units of resistivity are different as well.

$$R = \frac{\rho \ell}{A}$$

This equation only applies to shapes with constant cross sectional area, such as a cylinder or a rectangular prism, and will not apply to those with a varying cross sectional area, such as a sphere or a cone.

The resistivity of copper is around 10^{-8} Ω·m, while the resistivity of glass is *much* greater, around 10^{12} Ω·m.

Example 1 A wire of radius 1 mm and length 2 m is made of platinum (resistivity = 1×10^{-7} Ω·m). If a voltage of 9 V is applied between the ends of the wire, what will be the resulting current?

Solution. First, the resistance of the wire is given by the equation

$$R = \frac{\rho \ell}{A} = \frac{\rho \ell}{\pi r^2} = \frac{(1 \times 10^{-7} \; \Omega \cdot m)(2 \; m)}{\pi (0.001 \; m)^2} = 0.064$$

Then, from $I = V/R$, we get

$$I = \frac{V}{R} = \frac{9 \; V}{0.064} = 140 \; A$$

VOLTAGE

Now that we know how to measure current, the next question is, What causes it? Why does an electron drift through the circuit? One answer is to say that there's an electric field inside the wire, and since negative charges move in the direction opposite to the electric field lines, electrons would drift opposite the electric field.

Another (equivalent) answer to the question is that there's a difference in electric potential—otherwise known as a potential difference or, most simply, a voltage—between the ends of the wire. Negative charges naturally move from regions of higher potential to lower potential.

The River: Voltage
In terms of the river, a river will not flow if it is flat. A river will flow from higher ground to lower ground. We can think of Voltage as the mountain that gives the river height to flow down.

Voltage across two points along a conductor creates a current.

It is not uncommon to see the voltage that creates a current referred to as electromotive force (emf) ε, since it is the cause that sets the charges into motion in a preferred direction. The voltage is provided in a circuit by some source, which is typically a battery for direct current circuits.

Remember
Don't forget that emf is not a force. It is potential difference measured in volts which are J/C, not Newtons.

ELECTRIC CIRCUITS

An electric current is maintained when the terminals of a voltage source (a battery, for example) are connected by a conducting pathway, in what's called a **circuit**. If the current always travels in the same direction through the pathway, it's called a **direct current**.

To try to imagine what's happening in a circuit in which a steady-state current is maintained, let's follow one of the charge carriers drifting through the pathway. (Remember we're pretending that the charge carriers are positive.) The charge is introduced by the positive terminal of the battery and enters the wire, where it's pushed by the electric field. It encounters resistance, bumping into the relatively stationary atoms that make up the metal's lattice and setting them into greater motion. So the electrical potential energy that the charge had when it left the battery is turning into heat. By the time the charge reaches the negative terminal, all of its original electrical potential energy is lost. In order to keep the current going, the voltage source must do positive work on the charge, forcing it to move from the negative terminal toward the positive terminal. The charge is now ready to make another journey around the circuit.

Energy and Power

When a carrier of positive charge q drops by an amount V in potential, it loses potential energy in the amount qV. If this happens in time t, then the rate at which this energy is transformed is equal to $(qV)/t = (q/t)V$. But q/t is equal to the current, I, so the rate at which electrical energy is transferred is given by the equation

$$P = IV$$

Heat Dissipation
Power in a circuit is very closely associated with heat given off. If we were to touch a lightbulb we would notice it becomes hot. The greater the heat given off, the brighter the light bulb is.

Equation Sheet

This equation works for the power delivered by a battery to the circuit as well as for resistors. The power dissipated in a resistor, as electrical potential energy is turned into heat, is given by $P = IV$, but, because of the relationship $V = IR$, we can express this in two other ways:

$$P = IV = I(IR) = I^2R$$

or

$$P = IV = \frac{V}{R} \cdot V = \frac{V^2}{R}$$

Resistors become hot when current passes through them.

CIRCUIT ANALYSIS

We will now develop a way of specifying the current, voltage, and power associated with each element in a circuit. Our circuits will contain three basic elements: batteries, resistors, and connecting wires. As we've seen, the resistance of an ordinary metal wire is negligible; resistance is provided by devices that control the current: **resistors**. All the resistance of the system is concentrated in the resistors, which are symbolized in a circuit diagram by this symbol:

Batteries are denoted by the symbol:

where the longer line represents the **positive** (higher potential) terminal, and the shorter line is the **negative** (lower potential) terminal. Sometimes a battery is denoted by more than one pair of such lines:

Here's a simple circuit diagram:

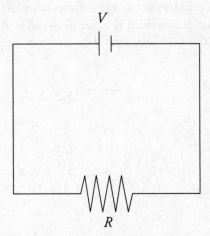

The electric potential (V) of the battery is indicated, as is the resistance (R) of the resistor. As stated before, before the electrons return to the battery, all their energy is lost in the circuit. Since the resistor is the only thing in this circuit, it has to have the same potential difference as the battery, just with the opposite sign. For now, this sign isn't important, but it will be later on when we cover Kirchhoff's rules. Determining the current in this case is straightforward. The equation $V = IR$ gives us

$$I = \frac{V}{R}$$

COMBINATIONS OF RESISTORS

Two common ways of combining resistors within a circuit is to place them either in **series** (one after the other),

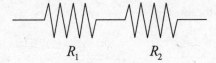

or in **parallel** (that is, side-by-side):

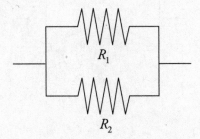

In order to simplify the circuit, our goal is to find the equivalent resistance of combinations. Resistors are said to be in series if they all share the same current and if the total voltage drop across them is equal to the sum of the individual voltage drops.

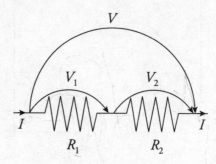

In this case, then, if V denotes the voltage drop across the combination, we have

$$R_{equiv} = \frac{V}{I} = \frac{V_1 + V_2}{I} = \frac{V_1}{I} + \frac{V_2}{I} = R_1 + R_2$$

This idea can be applied to any number of resistors in series (not just two):

Equation Sheet

$$R_S = \sum_i R_i$$

What's the same?
Remember, in a series circuit the current is the same through each resisitor—in parallel, the voltage drop is the same.

Which would you rather travel?
It makes sense that in a set of parallel resistors current will want to travel through the path of least resistance—if two resistors are in parallel and the first resistor has the greater resistance. The second resistor will have more of the current running through it.

Resistors are said to be in parallel if they all share the same voltage drop, and the total current entering the combination is split among the resistors. Imagine that a current I enters the combination. It splits; some of the current, I_1, would go through R_1, and the remainder, I_2, would go through R_2, such that the voltage is the same for each resistor.

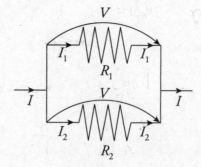

So if V is the voltage drop across the combination, we have

$$I = I_1 + I_2 \quad \Rightarrow \quad \frac{V}{R_{\text{equiv}}} = \frac{V}{R_1} + \frac{V}{R_2} \quad \Rightarrow \quad \frac{1}{R_{\text{equiv}}} = \frac{1}{R_1} + \frac{1}{R_2}$$

This idea can be applied to any number of resistors in parallel (not just two): The reciprocal of the equivalent resistance for resistors in parallel is equal to the sum of the reciprocals of the individual resistances:

$$\frac{1}{R_P} = \sum_i \frac{1}{R_i}$$

Pay Attention
A common mistake happens when calculating an equivalent resistor for a set of parallel resistors. Write out the equation. Students often forget to take the reciprocal as the final step in calculating the equivalent resistance.

Equation Sheet

Example 2 Calculate the equivalent resistance for the following circuit:

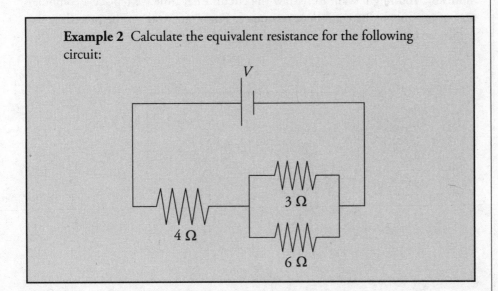

Solution. First find the equivalent resistance of the two parallel resistors:

$$\frac{1}{R_P} = \frac{1}{3\Omega} + \frac{1}{6\Omega} \quad \Rightarrow \quad \frac{1}{R_P} = \frac{1}{2\Omega} \quad \Rightarrow \quad R_P = 2$$

This resistance is in series with the 4 Ω resistor, so the overall equivalent resistance in the circuit is $R = 4\ \Omega + 2\ \Omega = 6\ \Omega$.

The reason for us considering these equivalent resistances is because the easiest problem to solve is the simple, one-resistor problem presented originally. You can represent many combinations of resistors in series or parallel as a single equivalent resistor, and then use Ohm's law to solve for the current.

Example 3 Determine the current through each resistor, the voltage drop across each resistor, and the power given off (dissipated) as heat in each resistor of this circuit:

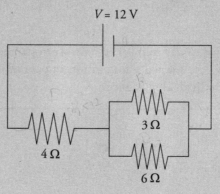

Solution. You might want to redraw the circuit each time we replace a combination of resistors by its equivalent resistance. From our work in the preceding example, we have

Make it simple then work backwards
The typical method to working a circuit problem is to make the circuit provided into a simple circuit with a battery and an equivalent resistor. Then to solve for individual values on each resistor, we work backwards and build the circuit back to its original form. In the diagram to the right we moved backwards and made the circuit simple from diagram 1 to 2 to 3. Then to solve for the resistors we go from 3 to 2 to 1.

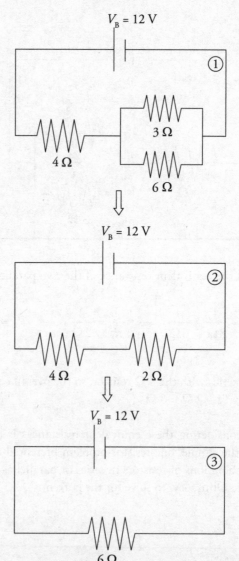

From diagram ③, which has just one resistor, we can figure out the current:

$$I = \frac{V_B}{R} = \frac{12\text{ V}}{6} = 2\text{ A}$$

Now we can work our way back to the original circuit (diagram ①). In going from ③ to ②, we are going back to a series combination, and what do resistors in series share? That's right, the same current. So, we take the current, $I = 2$ A, back to diagram ②. The current through each resistor in diagram ② is 2 A.

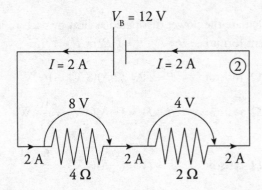

Since we know the current through each resistor, we can figure out the voltage drop across each resistor using the equation $V = IR$. The voltage drop across the 4 Ω resistor is (2 A)(4 Ω) = 8 V, and the voltage drop across the 2 Ω resistor is (2 A)(2 Ω) = 4 V. Notice that the total voltage drop across the two resistors is 8 V + 4 V = 12 V, which matches the emf of the battery.

Now for the last step; going from diagram ② back to diagram ①. Nothing needs to be done with the 4 Ω resistor; nothing about it changes in going from diagram ② to ①, but the 2 Ω resistor in diagram ② goes back to the parallel combination. And what do resistors in parallel share? The same voltage drop. So we take the voltage drop, $V = 4$ V, back to diagram ①. The voltage drop across each of the two parallel resistors in diagram ① is 4 V.

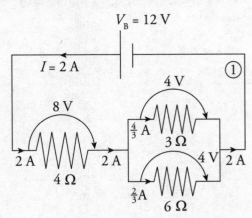

Since we know the voltage drop across each resistor, we can figure out the current through each resistor by using the equation $I = V/R$. The current through the 3 Ω resistor is $(4\ V)/(3\ Ω) = \dfrac{4}{3}$ A, and the current through the 6 Ω resistor is $(4\ V)/(6\ Ω) = \dfrac{2}{3}$ A. Note that the current entering the parallel combination (2 A) equals the total current passing through the individual resistors $\left(\dfrac{4}{3}\ A + \dfrac{2}{3}\ A\right)$. Again this was expected.

Finally, we will calculate the power dissipated as heat by each resistor. We can use any of the equivalent formulas: $P = IV$, $P = I^2R$, or $P = V^2/R$.

$\quad$ For the 4 Ω resistor: $\qquad P = IV = (2\ A)(8\ V) = 16\ W$

$\quad$ For the 3 Ω resistor: $\qquad P = IV = (\dfrac{4}{3}A)(4\ V) = \dfrac{16}{3}\ W$

$\quad$ For the 6 Ω resistor: $\qquad P = IV = (\dfrac{2}{3}A)(4\ V) = \dfrac{8}{3}\ W$

So, the resistors are dissipating a total of

$$16\ W + \dfrac{16}{3}\ W + \dfrac{8}{3}\ W = 24\ W$$

If the resistors are dissipating a total of 24 J every second, then they must be provided with that much power. This is easy to check: $P = IV = (2\ A)(12\ V) = 24\ W$.

Example 4 For the following circuit,
  (a) In which direction will current flow and why?
  (b) What's the overall emf?
  (c) What's the current in the circuit?
  (d) At what rate is energy consumed by, and provided to, this circuit?

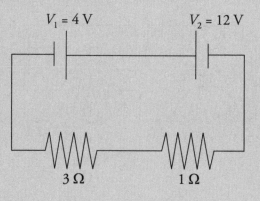

Solution.

(a) The battery V_1 wants to send current clockwise, while the battery V_2 wants to send current counterclockwise. Since $V_2 > V_1$, the battery whose emf is V_2 is the more powerful battery, so the current will flow counterclockwise.

(b) Charges forced through V_1 will lose, rather than gain, 4 V of potential, so the overall emf of this circuit is $V_2 - V_1 = 8$ V.

(c) Since the total resistance is $3\ \Omega + 1\ \Omega = 4\ \Omega$, the current will be $I = (8\text{ V})/(4\ \Omega) = 2$ A. Note that we use the overall emf, and not either of the individual battery emfs.

(d) V_2 will provide energy at a rate of $P_2 = IV_2 = (2\text{ A})(12\text{ V}) = 24$ W, while V_1 will absorb at a rate of $P_1 = IV_1 = (2\text{ A})(4\text{ V}) = 8$ W. Finally, energy will be dissipated in these resistors at a rate of $I^2R_1 + I^2R_2 = (2\text{ A})^2(3\ \Omega) + (2\text{ A})^2(1\ \Omega) = 16$ W. Once again, energy is conserved; the power delivered (24 W) equals the power taken (8 W + 16 W = 24 W).

Example 5 All real batteries contain **internal resistance**, r. Determine the current in the following circuit when the switch S is closed:

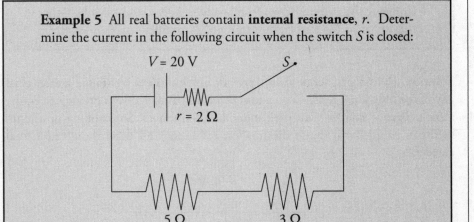

Solution. Before the switch is closed, there is no complete conducting pathway from the positive terminal of the battery to the negative terminal, so no current flows through the resistors. However, once the switch is closed, the resistance of the circuit is $2\ \Omega + 3\ \Omega + 5\ \Omega = 10\ \Omega$, so the current in the circuit is $I = (20\text{ V})/(10\ \Omega) = 2$ A. Often the battery and its internal resistance are enclosed in a dashed box:

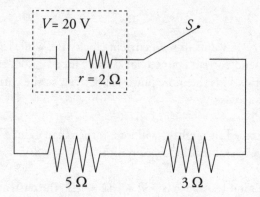

In this case, a distinction can be made between the emf of the battery and the actual voltage it provides once the current has begun. Since $I = 2$ A, the voltage drop across the internal resistance is $Ir = (2 \text{ A})(2 \text{ }\Omega) = 4$ V, so the effective voltage provided by the battery to the rest of the circuit—called the **terminal voltage**—is lower than the ideal voltage. It is $V = V_B - Ir = 20 \text{ V} - 4 \text{ V} = 16 \text{ V}$.

Example 6 A student has three 30 Ω resistors and an ideal 90 V battery. (A battery is *ideal* if it has a negligible internal resistance.) Compare the current drawn from—and the power supplied by—the battery when the resistors are arranged in parallel versus in series.

Solution. Resistors in series always provide an equivalent resistance greater than any of the individual resistances, and resistors in parallel always provide an equivalent resistance smaller than their individual resistances. So, hooking up the resistors in parallel will create the smallest resistance and draw the greatest total current:

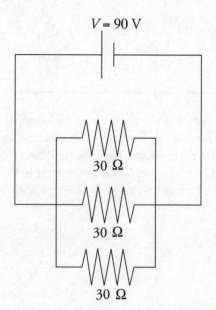

In this case, the equivalent resistance is

$$\frac{1}{R_P} = \frac{1}{30\ \Omega} + \frac{1}{30\ \Omega} + \frac{1}{30\ \Omega} \quad \Rightarrow \quad \frac{1}{R_P} = \frac{1}{10\ \Omega} \quad \Rightarrow \quad R_P = 10\ \Omega$$

and the total current is $I = V/R_p = (90\ \text{V})/(10\ \Omega) = 9$ A. (You could verify that 3 A of current would flow in each of the three branches of the combination.) The power supplied by the battery will be $P = IV = (9\ \text{A})(90\ \text{V}) = 810$ W.

If the resistors are in series, the equivalent resistance is $R_S = 30\ \Omega + 30\ \Omega + 30\ \Omega = 90\ \Omega$, and the current drawn is only $I = V/R_S = (90\ \text{V})/(90\ \Omega) = 1$ A. The power supplied by the battery in this case is just $P = IV = (1\ \text{A})(90\ \text{V}) = 90$ W.

Example 7 A **voltmeter** is a device used to measure the voltage between two points in a circuit. An **ammeter** is used to measure current. Determine the readings on the voltmeter (denoted —Ⓥ—) and the ammeter (denoted —Ⓐ—) in the circuit below.

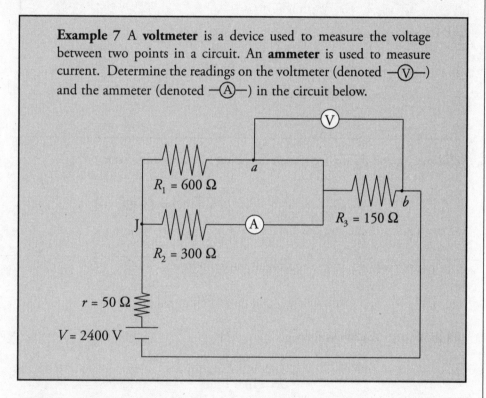

Solution. We consider the ammeter to be ideal; this means it has negligible resistance, so it doesn't alter the current that it's trying to measure. Similarly, we consider the voltmeter to have an extremely high resistance, so it draws negligible current away from the circuit.

Our first goal is to find the equivalent resistance in the circuit. The 600 Ω and 300 Ω resistors are in parallel; they're equivalent to a single 200 Ω resistor. This is in series with the battery's internal resistance, r, and R_3. The overall equivalent resistance is therefore $R = 50\ \Omega + 200\ \Omega + 150\ \Omega = 400\ \Omega$, so the current supplied by the battery is $I = V/R (2400\ \text{V})/(400\ \Omega) = 6$ A. At the junction marked J, this current splits. Since R_1 is twice R_2, half as much current will flow through R_1 as through R_2; the current through R_1 is $I_1 = 2$ A, and the current through R_2 is $I_2 = 4$ A. The voltage drop across each of these resistors is $I_1 R_1 = I_2 R_2 = 1{,}200$ V (matching voltages verify the values of currents I_1 and I_2). Since the ammeter is in the branch that contains R_2, it will read $I_2 = 4$ A.

The voltmeter will read the voltage drop across R_3, which is $V_3 = IR_3 = (6$ A$)$ $(150\ \Omega) = 900$ V. So the potential at point b is 900 V lower than at point a.

Example 8 The diagram below shows a point a at potential $V = 20$ V connected by a combination of resistors to a point (denoted G) that is **grounded**. *The ground is considered to be at potential zero.* If the potential at point a is maintained at 20 V, what is the current through R_3?

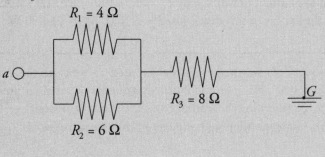

Solution. R_1 and R_2 are in parallel; their equivalent resistance is R_p, where

$$\frac{1}{R_p} = \frac{1}{4\ \Omega} + \frac{1}{6\ \Omega} \quad \Rightarrow \quad R_p = \frac{24}{10}\ \Omega = 2.4\ \Omega$$

R_p is in series with R_3, so the equivalent resistance is:

$$R = R_p + R_3 = (2.4\ \Omega) + (8\ \Omega) = 10.4\ \Omega$$

and the current that flows through R_3 is

$$I_3 = \frac{V}{R} = \frac{20\ \text{V}}{10.4\ \Omega} = 1.9\ \text{A}$$

Kirchhoff's Rules

When the resistors in a circuit cannot be classified as either in series or in parallel, we need another method for analyzing the circuit. The rules of Gustav Kirchhoff (pronounced "Keer koff") can be applied to any circuit:

The Loop Rule. The sum of the potential differences (positive and negative) that traverse any closed loop in a circuit must be zero.

The Junction Rule. The total current that enters a junction must equal the total current that leaves the junction. (This is also known as the **Node Rule**.)

The Loop Rule just says that, starting at any point, by the time we get back to that same point by following any closed loop, we have to be back to the same potential. Therefore, the total drop in potential must equal the total rise in potential. Put another way, the Loop Rule says that all the decreases in electrical potential energy (for example, caused by resistors in the direction of the current) must be balanced by all the increases in electrical potential energy (for example, caused by a source of emf from the negative to positive terminal). So the Loop Rule is basically a re-statement of the Law of Conservation of Energy.

Similarly, the Junction Rule simply says that the charge (per unit time) that goes into a junction must equal the charge (per unit time) that comes out. This is basically a statement of the Law of Conservation of Charge. In practice, the Junction Rule is straightforward to apply.

The most important things to remember about the Loop Rule can be summarized as follows:

- First, you choose a direction for the loop—either clockwise or counterclockwise, it doesn't matter.
- When the loop goes across a resistor in the *same* direction as the current, the potential *drops* by *IR*.
- When the loop goes across a resistor in the *opposite* direction from the current, the potential *increases* by *IR*.
- When the loop goes from the negative to the positive terminal of a source of emf, the potential *increases* by *V*.
- When the loop goes from the positive to the negative terminal of a source of emf, the potential *decreases* by *V*.

What goes in must come out
Knowing that the total drop must equal the total rise in potential is a good way to keep yourself in check when doing circuit problems. If 60 V came out of a battery by the end no more and no less must be used.

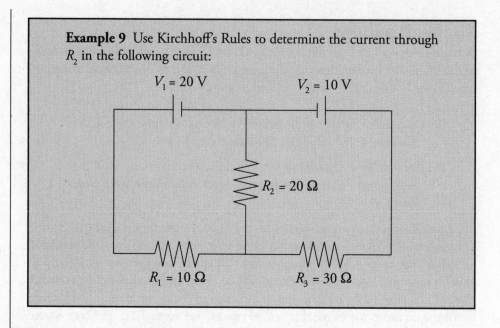

Example 9 Use Kirchhoff's Rules to determine the current through R_2 in the following circuit:

$V_1 = 20$ V

$V_2 = 10$ V

$R_2 = 20 \ \Omega$

$R_1 = 10 \ \Omega$

$R_3 = 30 \ \Omega$

Solution. First let's label some points in the circuit.

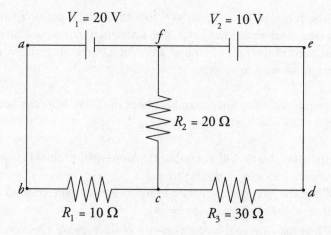

$V_1 = 20$ V

$V_2 = 10$ V

$R_2 = 20 \ \Omega$

$R_1 = 10 \ \Omega$

$R_3 = 30 \ \Omega$

The points c and f are junctions (nodes). We have two nodes and three branches: one branch is *fabc*, another branch is *cdef*, and the third branch is *cf*. Each branch has one current throughout. If we label the current in *fabc* I_1 and the current in branch *cdef* I_2 (with the directions as shown in the diagram below), then the current in branch *cf* must be $I_1 - I_2$, by the Junction Rule: I_1 comes into c, and a total of $I_2 + (I_1 - I_2) = I_1$ comes out.

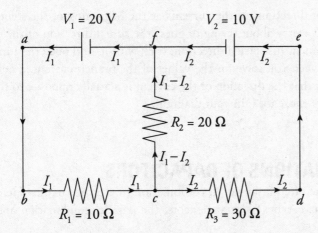

Now pick a loop, say, *abcfa*, in the counterclockwise direction. Starting at *a*, we go to *b*, then across R_1 in the direction of the current, so the potential drops by $I_1 R_1$. Then we move to *c*, then up through R_2 in the direction of the current, so the potential drops by $(I_1 - I_2)R_2$. Then we reach *f*, turn left and travel through V_1 from the negative to the positive terminal, so the potential increases by V_1. We now find ourselves back at *a*. By the Loop Rule, the total change in potential around this closed loop must be zero, and

$$-I_1 R_1 - (I_1 - I_2)R_2 + V_1 = 0$$

Since we have two unknowns (I_1 and I_2), we need two equations, so now pick another loop; let's choose *cdefc*, again in the counterclockwise direction. From *c* to *d*, we travel across the resistor in the direction of the current, so the potential drops by $I_2 R_3$. From *e* to *f*, we travel through V_2 from the positive to the negative terminal, so the potential *drops* by V_2. Heading down from *f* to *c*, we travel across R_2 but in the direction *opposite* to the current, so the potential *increases* by $(I_1 - I_2)R_2$. At *c*, our loop is completed, so

$$-I_2 R_3 - V_2 + (I_1 - I_2)R_2 = 0$$

Substituting in the given numerical values for R_1, R_2, R_3, V_1, and V_2, and simplifying, these two equations become

$$3I_1 - 2I_2 = 2$$
$$2I_1 - 5I_2 = 1$$

Solving this pair of simultaneous equations, we get

$$I_1 = \frac{8}{11} \text{ A} = 0.73 \text{ A} \quad \text{and} \quad I_2 = \frac{1}{11} \text{ A} = 0.09 \text{ A}$$

So the current through R_2 is $I_1 - I_2 = \frac{7}{11}$ A = 0.64 A.

The choice of directions of the currents at the beginning of the solution was arbitrary. Don't worry about trying to guess the actual direction of the current in a particular branch. Just pick a direction, stick with it, and obey the Junction Rule. At the end, when you solve for the values of the branch current, a negative value will alert you that the direction of the current is actually opposite to the direction you originally chose for it in your diagram.

COMBINATIONS OF CAPACITORS

Capacitors are often arranged in combination in electric circuits. Here we'll look at the same two types of arrangements, the parallel combination and the series combination.

A collection of capacitors are said to be in **parallel** if they all share the same potential difference. The following diagram shows two capacitors wired in parallel:

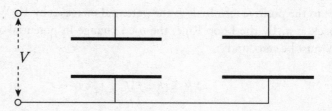

The top plates are connected by a wire and form a single equipotential; the same is true for the bottom plates. Therefore, the potential difference across one capacitor is the same as the potential difference across the other capacitor.

We want to find the capacitance of a *single* capacitor that would perform the same function as this combination. If the capacitances are C_1 and C_2, then the charge on the first capacitor is $Q_1 = C_1V$ and the charge on the second capacitor is $Q_2 = C_2V$. The total charge on the combination is $Q_1 + Q_2$, so the equivalent capacitance, C_P, must be

Equation Sheet

$$C_P = \frac{Q}{V} = \frac{Q_1 + Q_2}{V} = \frac{Q_1}{V} + \frac{Q_2}{V} = C_1 + C_2$$

So the **equivalent** capacitance of a collection of capacitors in parallel is found by adding the individual capacitances.

A collection of capacitors are said to be in **series** if they all share the same charge magnitude. This is similar to resistors in series, except resistors in series would share current, not charge. The following diagram shows two capacitors wired in series:

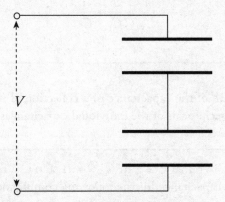

When a potential difference is applied, as shown, negative charge will be deposited on the bottom plate of the bottom capacitor; this will push an equal amount of negative charge away from the top plate of the bottom capacitor toward the bottom plate of the top capacitor. When the system has reached equilibrium, the charges on all the plates will have the same magnitude:

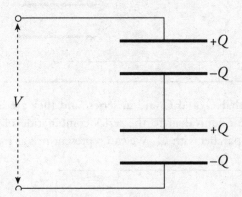

If the top and bottom capacitors have capacitances of C_1 and C_2, respectively, then the potential difference across the top capacitor is $V_1 = Q/C_1$, and the potential difference across the bottom capacitor is $V_2 = Q/C_2$. The total potential difference across the combination is $V_1 + V_2$, which must equal V. Therefore, the equivalent capacitance, C_S, must be

$$C_S = \frac{Q}{V} = \frac{Q}{V_1 + V_2} = \frac{Q}{\dfrac{Q}{C_1} + \dfrac{Q}{C_2}} = \frac{1}{\dfrac{1}{C_1} + \dfrac{1}{C_2}}$$

We can write this in another form:

Equation Sheet

$$\frac{1}{C_S} = \frac{1}{C_1} + \frac{1}{C_2}$$

In words, the *reciprocal* of the capacitance of a collection of capacitors in series is found by adding the reciprocals of the individual capacitances.

Example 10 Given that $C_1 = 2 \text{ μF}$, $C_2 = 4 \text{ μF}$, and $C_3 = 6 \text{ μF}$, calculate the equivalent capacitance for the following combination:

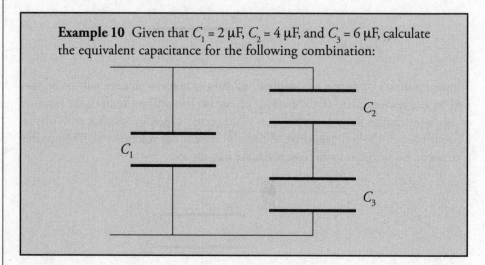

Solution. Notice that C_2 and C_3 are in series, and they are in parallel with C_1. That is, the capacitor equivalent to the series combination of C_2 and C_3 (which we'll call C_{2-3}) is in parallel with C_1. We can represent this as follows:

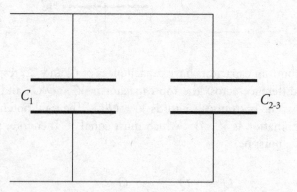

So, the first step is to find C_{2-3}:

$$\frac{1}{C_{2-3}} = \frac{1}{C_2} + \frac{1}{C_3} \quad \Rightarrow \quad C_{2-3} = \frac{C_2 C_3}{C_2 + C_3}$$

Now this is in parallel with C_1, so the overall equivalent capacitance ($C_{1\text{-}2\text{-}3}$) is

$$C_{1\text{-}2\text{-}3} = C_1 + C_{2\text{-}3} = C_1 + \frac{C_2 C_3}{C_2 + C_3}$$

Substituting in the given numerical values, we get

$$C_{1\text{-}2\text{-}3} = (2\mu\ \text{F}) + \frac{(4\mu\ \text{F})(6\mu\ \text{F})}{(4\mu\ \text{F}) + (6\mu\ \text{F})} = 4.4\mu\ \text{F}$$

CAPACITORS AND DIELECTRICS REVISITED

Now let's examine the case in which a capacitor is first charged up and, while it's still connected to its voltage source, has a dielectric inserted between its plates. First, since the capacitor is still connected to the battery, the voltage between the plates must match the voltage of the battery. Therefore, V will not change. Next, because the capacitance C increases, the equation $Q = CV$ tells us that the charge Q must increase; in fact, because V doesn't change and C increases by a factor of κ, we see that Q will increase by a factor of κ. Next, using the equation $V = Ed$, we see that because V doesn't change, neither will E. Finally, using the equation $U = \frac{1}{2}QV$, we conclude that since V doesn't change and Q increases by a factor of κ, the stored electric potential energy increases by a factor of κ. An important point to notice is that V doesn't change because the battery will transfer additional charge to the capacitor plates. This increase in Q offsets any momentary decrease in the electric field strength when the dielectric is inserted (because the molecules of the dielectric are polarized, as above) and brings the electric field strength back to its original value. Furthermore, as more charge is transferred to the plates, more electric potential energy is stored.

The following figures summarize the effects on the properties of a capacitor with a dielectric inserted between the plates.

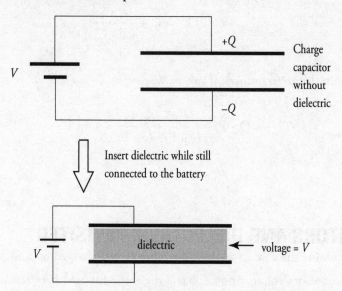

C increases by a factor of κ	C always increases
Q increases by a factor of κ	battery disconnected
V stays the same	$Q = \uparrow C \downarrow V$
E stays the same	$\downarrow V = \downarrow Ed$
U_{PE} increases by a factor of κ	$\downarrow U = (1/2) Q \downarrow V$

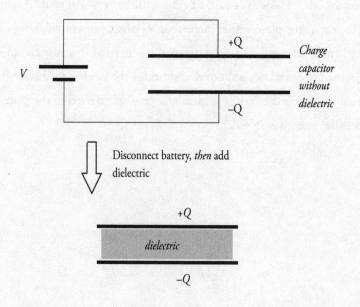

C increases by a factor of κ	C always increases
Q stays the same	battery disconnected
V decreases by a factor of κ	$\uparrow Q = \uparrow C\ V$
E decreases by a factor of κ	$V = E\,d$
U_{PE} decreases by a factor of κ	$\uparrow U = (1/2)\uparrow Q\,V$

RESISTANCE–CAPACITANCE (RC) CIRCUITS

Capacitors are typically charged by batteries. Once the switch in the diagram on the left is closed, electrons are attracted to the positive terminal of the battery and leave the top plate of the capacitor. Electrons also accumulate on the bottom plate of the capacitor, and this continues until the voltage across the capacitor plates matches the emf of the battery. When this condition is reached, the current stops and the capacitor is fully charged.

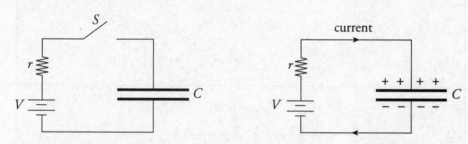

Example 11 Determine the current through and the voltage across each electrical device in the following circuit when

    (a) the switch has just been closed

    (b) the switch has been thrown for a long period of time

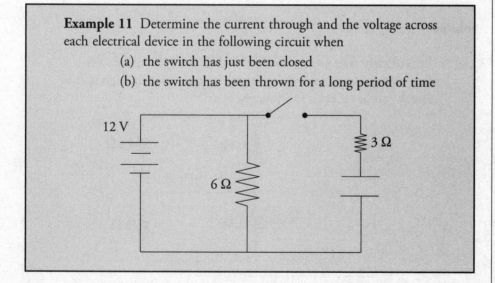

Solution.

(a) When the switch has just been closed, the capacitor is treated like a wire with no resistance. The circuit behaves like a simple parallel circuit. There are 12 V across the 6 Ω resistor so 2 amps of current flow through it. There are 12 V across the 3 Ω resistor so 4 amps flow through it. By Kirchhoff's Law, the currents add to 6 amps through the battery and the equivalent resistance of the circuit is 2 Ω.

(b) After the switch has been closed for a long period of time, the capacitor behaves like an infinite resistor. There are still 12 V across the 6 Ω resistor so 2 amps of current still flow through it. If the capacitor behaves like an infinite resistor, no current can flow through that branch. There will be no current through the 3 Ω resistor. The capacitor will have 12 V across it and there will be no voltage across the 3 Ω resistor (otherwise current would be flowing through it). The current will be 2 amps through the battery and the equivalent resistance of the circuit is 6 Ω.

Example 12 Determine the current through and the voltage across each electrical device in the following circuit when

(a) the switch has just been closed

(b) the switch has been thrown for a long period of time

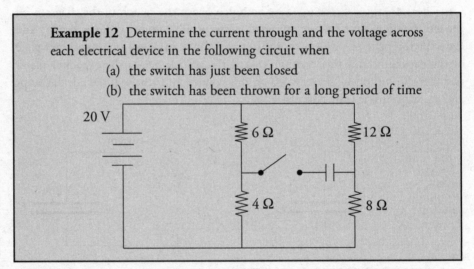

Solution.

(a) Immediately after one throws the switch, the capacitor exhibits behavior similar to that of a wire with no resistance through it. The circuit can be redrawn and thought of as

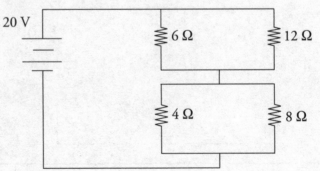

The 6 Ω and 12 Ω resistors are in parallel and can be thought of as 4 Ω (from $1/R_{eq} = 1/R_1 + 1/R_2$). Similarly, the 4 Ω and 8 Ω resistors are in parallel and can be thought of as 2.67 Ω (again, from $1/R_{eq} = 1/R_1 + 1/R_2$). The equivalent resistance of these two branches, which are in series with each other, is 6.67 Ω (from $R_{eq} = R_1 + R_2$).

The current through the battery is therefore 3 amps (from Ohm's Law, $V = IR$).

The current through the 6 Ω and 12 Ω (top two resistors) is 3 amps traveling through an effective resistance of 4 Ω. The voltage drop across the top branch is 12 V (from Ohm's Law). The currents through the 6 Ω and 12 Ω resistors are 2 amps and 1 amp, respectively (again from Ohm's Law).

The current through the 4 Ω and 8 Ω (bottom two resistors) is also 3 amps through an effective resistance of 2.67 Ω. The voltage drop across the branch is 8 V (from Ohm's Law or from the fact that the voltage across each series section must sum to the total voltage across the battery). The currents through the 4 Ω and 8 Ω resistors are 2 amps and 1 amp, respectively (again from Ohm's Law).

(b)   When the switch has been closed for a long period of time, the capacitor behaves just like an infinite resistor. The circuit can be redrawn and thought of as

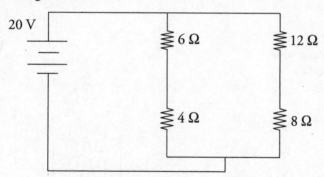

The 6 Ω and 4 Ω resistors are in series and can be thought of as a 10 Ω resistor (from $R_{eq} = R_1 + R_2$). Similarly, the 12 Ω and 8 Ω resistors can be thought of as 20 Ω (again, from $R_{eq} = R_1 + R_2$). Since the branches are in parallel, the voltage is 20 volts across each path and the current can be calculated by Ohm's Law ($V = IR$); 2 amps go through the left branch (the 6 Ω and 4 Ω resistor) and 1 amp goes through the right branch (the 12 Ω and 8 Ω resistor).

The current through the battery is therefore 3 amps (Kirchhoff's Node Rule).

The voltage across each of the 6 Ω, 4 Ω, 12 Ω, and 8 Ω resistors can be obtained from Ohm's Law. They are 12 V, 8 V, 12 V, and 8 V, respectively.

Chapter 7 Review Questions

Solutions can be found in Chapter 12.

Section I: Multiple Choice

1. A wire made of brass and a wire made of silver have the same length, but the diameter of the brass wire is 4 times the diameter of the silver wire. The resistivity of brass is 5 times greater than the resistivity of silver. If R_B denotes the resistance of the brass wire and R_S denotes the resistance of the silver wire, which of the following is true?

   (A) $R_B = \dfrac{5}{16} R_S$

   (B) $R_B = \dfrac{4}{5} R_S$

   (C) $R_B = \dfrac{5}{4} R_S$

   (D) $R_B = \dfrac{5}{2} R_S$

2. For an ohmic conductor, doubling the voltage without changing the resistance will cause the current to

   (A) decrease by a factor of 4
   (B) decrease by a factor of 2
   (C) increase by a factor of 2
   (D) increase by a factor of 4

3. If a 60-watt light bulb operates at a voltage of 120 V, what is the resistance of the bulb?

   (A) $2 \, \Omega$
   (B) $30 \, \Omega$
   (C) $240 \, \Omega$
   (D) $720 \, \Omega$

4. A battery whose emf is 40 V has an internal resistance of $5 \, \Omega$. If this battery is connected to a $15 \, \Omega$ resistor R, what will the voltage drop across R be?

   (A) 10 V
   (B) 30 V
   (C) 40 V
   (D) 50 V

5.

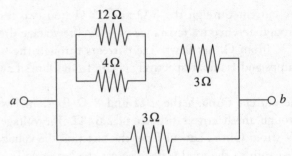

   Determine the equivalent resistance between points a and b.

   (A) $0.25 \, \Omega$
   (B) $0.333 \, \Omega$
   (C) $1.5 \, \Omega$
   (D) $2 \, \Omega$

6.

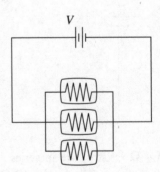

   Three identical light bulbs are connected to a source of emf, as shown in the diagram above. What will happen if the middle bulb burns out?

   (A) The light intensity of the other two bulbs will decrease (but they won't go out).
   (B) The light intensity of the other two bulbs will increase.
   (C) The light intensity of the other two bulbs will remain the same.
   (D) More current will be drawn from the source of emf.

7.

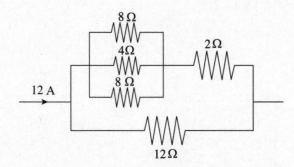

What is the voltage drop across the 12 Ω resistor in the portion of the circuit shown above?

(A) 24 V
(B) 36 V
(C) 48 V
(D 72 V

8.

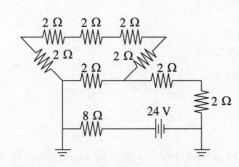

What is the current through the 8 Ω resistor in the circuit shown above?

(A) 0.5 A
(B) 1.0 A
(C) 1.5 A
(D) 3.0 A

9. How much energy is dissipated as heat in 20 s by a 100 Ω resistor that carries a current of 0.5 A?

(A) 50 J
(B) 100 J
(C) 250 J
(D) 500 J

10.

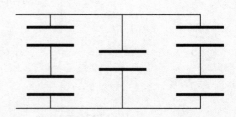

If each of the capacitors in the array shown above is C, what is the capacitance of the entire combination?

(A) C/2
(B) 2C/3
(C) 5C/6
(D) 2C

Section II: Free Response

1. Consider the following circuit:

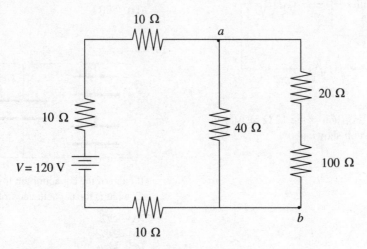

(a)   At what rate does the battery deliver energy to the circuit?

(b)   Find the current through the 40 Ω resistor.

(c)   (i)   Determine the potential difference between points *a* and *b*.

       (ii)   At which of these two points is the potential higher?

(d)   Find the energy dissipated by the 100 Ω resistor in 10 s.

(e)   Given that the 100 Ω resistor is a solid cylinder that's 4 cm long, composed of a material whose resistivity is 0.45 Ω·m, determine its radius.

2. Consider the following circuit:

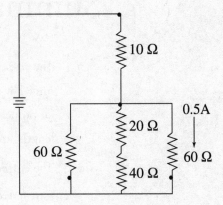

(a)   What is the current through each resistor?

(b)   What is the potential difference across each resistor?

(c)   What is the equivalent resistance of the circuit?

Summary

o The resistance of an object can be determined by $R = \dfrac{\rho \ell}{A}$ where ρ is the resistivity (a property of the material), ℓ is the length, and A is the cross-sectional area.

o The current is the rate at which charge is transferred and given by $I_{AVG} = \dfrac{\Delta Q}{\Delta t}$.

o Many objects obey Ohm's Law, which is given by $V = IR$.

o The electrical power in a circuit is given by $P = IV$ or $P = I^2R$ or $P = \dfrac{V^2}{R}$.

This is the same power we've encountered in our discussion of energy $P = \dfrac{W}{t}$.

o
In a series circuit	In a parallel circuit
$V_B = V_1 + V_2 + \dots$	$V_B = V_1 = V_2 = \dots$
$I_B = I_1 = I_2 = \dots$	$I_B = I_1 + I_2 + \dots$
$R_{eq} = R_1 + R_2 + \dots$	$1/R_{eq} = 1/R_1 + 1/R_2 + \dots$
$R_S = \sum_i R_i$	$\dfrac{1}{R_P} = \sum_i \dfrac{1}{R_i}$

o Kirchhoff's Loop Rule tells us that the sum of the potential differences in any closed loop in a circuit must be zero.

o Kirchhoff's Junction Rule (Node Rule) tells us the total current that enters a junction must equal the total current that leaves the junction.

o To find the combined capacitance of capacitors in parallel (side by side), simply add their capacitances.

$$C_P = C_1 + C_2 + \dots \text{ or } C_P = \sum_i C_i.$$

o To find the capacitance of capacitors in series (one after another), add their inverses:

$$\frac{1}{C_S} = \frac{1}{C_1} + \frac{1}{C_2} + \dots \text{ or } \frac{1}{C_S} = \sum_i \frac{1}{C_i}.$$

Chapter 8
Magnetic Forces and Fields

INTRODUCTION

In Chapter 5, we learned that electric charges are the sources of electric fields and that other charges experience an electric force in those fields. The charges generating the field were assumed to be at rest, because if they weren't, then another force field would have been generated in addition to the electric field. Electric charges *that move* are the sources of **magnetic fields**, and other charges that move can experience a magnetic force in these fields.

MAGNETIC FIELDS

Similar to our discussion about electric fields, the space surrounding a magnet is permeated by a magnetic field. The direction of the magnetic field is defined as pointing out of the north end of a magnet and into the south end of a magnet as illustrated below.

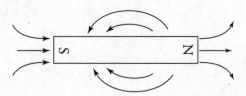

When two magnets get near each other the magnetic fields interfere with each other and can be drawn as follows. Note that, for simplicity's sake, only the field lines closest to the poles are shown.

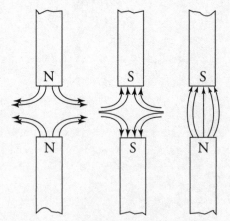

Notice there is a curve to the above fields. We call the field uniform if the field lines are parallel and of equal strength. It is easy to recognize a uniform magnetic field to the right, left, top of the page, or bottom of the page. But you will also see a field going into or out of the page. A field into the page looks as if there were a north pole of a magnet above the page pointing down at the south pole of a magnet that is below the page. It is represented by an area with X's going into the page. A field coming out of the page looks as if there were a north pole of a magnet below the page pointing up at the south pole of a magnet that is above the page. It is represented by an area with dots (•) coming out of the page.

INTO	OUT OF
X X X X	• • • •
X X X X	• • • •
X X X X	• • • •
X X X X	• • • •

BAR MAGNETS

A permanent bar magnet creates a magnetic field that closely resembles the magnetic field produced by a circular loop of current-carrying wire:

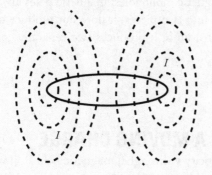 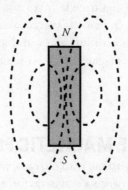

By convention, the magnetic field lines emanate from the end of the magnet designated the north pole (N) and then curl around and re-enter the magnet at the end designated the south pole (S). The magnetic field decreases in strength as distance from the magnet increases. The magnetic field created by a permanent bar magnet is due to the electrons, which have an intrinsic spin as they orbit the nuclei. They are literally charges in motion, which is the ultimate source of any and all magnetic fields. If a piece of iron is placed in an external magnetic field (for example, one created by a current-carrying solenoid), the individual magnetic dipole moments of the electrons will be forced to more or less line up. Because iron is ferromagnetic, these now-aligned magnetic dipole moments tend to retain this configuration, thus permanently magnetizing the bar and causing it to produce its own magnetic field.

As with electric charges, like magnetic poles repel each other, while opposite magnetic poles attract each other.

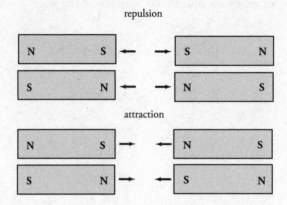

However, while you can have a positive electric charge all by itself, you can't have a single magnetic pole all by itself: the existence of a lone magnetic pole has never been confirmed. That is, there are no magnetic monopoles; magnetic poles always exist in pairs. If you break a bar magnet into two pieces, it does not produce one piece with just an N and another with just an S; it produces two separate, complete magnets, each with an N-S pair.

THE MAGNETIC FORCE ON A MOVING CHARGE

If a particle with charge q moves with velocity $\mathbf{v}$ through a magnetic field $\mathbf{B}$, it will experience a magnetic force, $\mathbf{F}_B$:

Equation Sheet

$$\mathbf{F}_B = q\mathbf{v} \times \mathbf{B}$$

with magnitude:

Equation Sheet

$$F_B = |q|\,vB\sin\theta$$

where θ is the angle between $\mathbf{v}$ and $\mathbf{B}$. From this equation, we can see that if the charge is at rest, then $v = 0$ immediately gives us $F_B = 0$. This tells us that magnetic forces only act on moving charges. Also, if $\mathbf{v}$ is parallel (or antiparallel) to $\mathbf{B}$, then $F_B = 0$ since, in either of these cases, $\sin\theta = 0$. So, only charges that cut across the magnetic field lines will experience a magnetic force. Furthermore, the magnetic force is maximized when **v** is perpendicular to **B**, since if $\theta = 90°$, then $\sin \theta$ is equal to 1, its maximum value.

The direction of $\mathbf{F}_B$ is always perpendicular to both **v** and **B** and depends on the sign of the charge q and the direction of $\mathbf{v} \times \mathbf{B}$ (which can be found by using the right-hand rule).

If q is *positive*, use your *right* hand and the *right*-hand rule.

If q is *negative*, you need your thumb to point in the direction of the product of q and v, not just the direction of the velocity.

Whenever you use the right-hand rule, follow these steps:

1. Orient your hand so that your thumb points in the direction of the velocity **v**. If the charge is negative, turn your thumb by 180 degrees.
2. Point your fingers in the direction of **B**.
3. The direction of $\mathbf{F}_B$ will then be perpendicular to your palm.

Think of your palm pushing with the force $\mathbf{F}_B$; the direction it pushes is the direction of $\mathbf{F}_B$.

Right-Hand Rule:

For determining the direction of the magnetic force, $\mathbf{F}_B$, on a *positive* charge

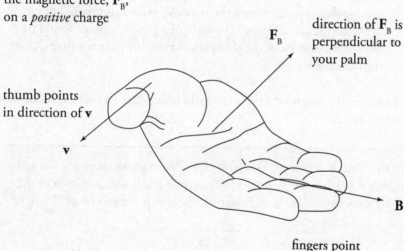

direction of $\mathbf{F}_B$ is perpendicular to your palm

thumb points in direction of **v**

fingers point in direction of **B**

Note that there are fundamental differences between the electric force and magnetic force on a charge. First, a magnetic force acts on a charge only if the charge is moving; the electric force acts on a charge whether it moves or not. Second, the direction of the magnetic force is always perpendicular to the magnetic field, while the electric force is always parallel (or antiparallel) to the electric field.

0 or 180
If the velocity v and the magnetic field **B** are parallel or anti-parallel, which means oriented so the angle between the vectors is 180 degrees, the magnetic force $\mathbf{F}_B = 0$.

Example 1 For each of the following charged particles moving through a magnetic field, determine the direction of the force acting on the charge.

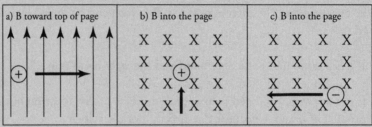

| a) B toward top of page | b) B into the page | c) B into the page |

Solution.

a) If you point your fingers to the top of the page and thumb to the right of the page, your palm should point out of the page. The force is out of the page.

b) If you point your fingers into the page and thumb to the top of the page your palm should point to the left of the page. The force is to the left of the page.

c) If you point your fingers into the page and thumb to the right of the page (remember, when the charge is negative, your thumb points in the opposite direction from the velocity), your palm should point to the bottom of the page. The force of a positive charge would be toward the bottom of the page, but because this is a negative charge, the force points up.

The SI unit for the magnetic field is the **tesla** (abbreviated **T**) which is one newton per ampere-meter.

Example 2 A charge $+q = +6 \times 10^{-6}$ C moves with speed $v = 4 \times 10^5$ m/s through a magnetic field of strength $B = 0.4$ T, as shown in the figure below. What is the magnetic force experienced by q?

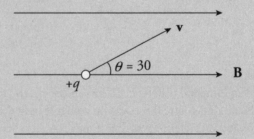

Solution. The magnitude of F_B is

$$F_B = qvB \sin \theta = (6 \times 10^{-6} \text{ C})(4 \times 10^5 \text{ m/s})(0.4 \text{ T}) \sin 30° = 0.48 \text{ N}$$

By the right-hand rule, the direction is into the plane of the page, which is symbolized by ×.

> **Example 3** A particle of mass m and charge $+q$ is projected with velocity **v** (in the plane of the page) into a uniform magnetic field **B** that points into the page. How will the particle move?
>
> $$\begin{matrix} \times & \times & \times & \times & \times \\ \mathbf{B} & & & & \\ \times & \times & \times \uparrow^{\mathbf{v}} & \times & \times \\ & & \bigcirc^{+q} & & \\ \times & \times & \times & \times & \times \\ & & & & \\ \times & \times & \times & \times & \times \end{matrix}$$

Solution. Since v is perpendicular to B, the particle will feel a magnetic force of strength qvB, which will be directed perpendicular to v (and to B) as shown:

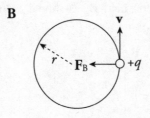

Since F_B is always perpendicular to **v**, the particle will undergo uniform circular motion; F_B will provide the centripetal force. Notice that, because F_B is always perpendicular to **v**, the magnitude of **v** will not change, just its direction. *Magnetic forces alone cannot change the speed of a charged particle, they can only change its direction of motion.* The radius of the particle's circular path is found from the equation $F_B = F_C$:

$$qvB = \frac{mv^2}{r} \implies r = \frac{mv}{qB}$$

Magnetic Fields
F_B is always perpendicular to both **v** and **B**. Magnetic forces cannot change the speed of an object, only its direction. The magnetic field does no work on any charges.

Example 4 A particle of charge $-q$ is shot into a region that contains an electric field, **E**, crossed with a perpendicular magnetic field, **B**. If $E = 2 \times 10^4$ N/C and $B = 0.5$ T, what must be the speed of the particle if it is to cross this region without being deflected?

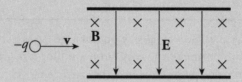

Solution. If the particle is to pass through undeflected, the electric force it feels has to be canceled by the magnetic force. In the diagram above, the electric force on the particle is directed upward (since the charge is negative and E is downward), and the magnetic force is directed downward by the right-hand rule. So F_E and F_B point in opposite directions, and in order for their magnitudes to balance, qE must equal qvB, so v must equal E/B, which in this case gives

$$v = \frac{E}{B} = \frac{2 \times 10^4 \text{ N/C}}{0.5 \text{ T}} = 4 \times 10^4 \text{ m/s}$$

Example 5 A particle with charge $+q$, traveling with velocity **v**, enters a uniform magnetic field **B**, as shown below. Describe the particle's subsequent motion.

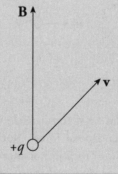

Solution. If the particle's velocity were parallel to **B**, then it would be unaffected by **B**. If **v** were perpendicular to **B**, then it would undergo uniform circular motion (as we saw in Example 2). In this case, **v** is neither purely parallel nor perpendicular to **B**. It has a component ($\mathbf{v}_1$) that's parallel to **B** and a component ($\mathbf{v}_2$) that's perpendicular to **B**.

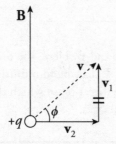

Component $\mathbf{v}_1$ will not be changed by **B**, so the particle will continue upward in the direction of **B**. However, the presence of $\mathbf{v}_2$ will create circular motion. The superposition of these two types of motion will cause the particle's trajectory to be a *helix*; it will spin in circular motion while traveling upward with the speed $v_1 = v \sin \phi$:

THE MAGNETIC FORCE ON A CURRENT-CARRYING WIRE

Since magnetic fields affect moving charges, they should also affect current-carrying wires. After all, a wire that contains a current contains charges that move. Remember, a current is the flow of positive charges.

Let a wire of length ℓ be immersed in magnetic field **B**. If the wire carries a current I, then the force on the wire is

$$\mathbf{F}_{\text{B}} = I\boldsymbol{\ell} \times \mathbf{B}$$

Equation Sheet

with magnitude

$$F_B = BI\ell \sin \theta$$

Equation Sheet

Another Way of Seeing It
Current I is charge over time (q/t) and length of a wire l is a distance d. Hence I ℓ = q (d/t). This gives us the same formula as before:
F_B = qvbsin θ = I ℓ Bsin θ

where θ is the angle between ℓ and **B**. Here, the direction of ℓ is the direction of the current, I. The direction of $\mathbf{F}_B$ can be found using the right-hand rule and by letting your thumb point in the direction in which the current flows. Remember, a current is the flow of positive charges.

Example 6 A U-shaped wire of mass m is lowered into a magnetic field **B** that points out of the plane of the page. How much current I must pass through the wire in order to cause the net force on the wire to be zero?

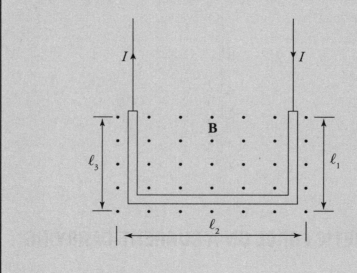

Solution. The total magnetic force on the wire is equal to the sum of the magnetic forces on each of the three sections of wire. The force on the first section (the right, vertical one), $\mathbf{F}_{B1}$, is directed to the left (applying the right-hand rule, and the force on the third piece (the left, vertical one), $\mathbf{F}_{B3}$, is directed to the right. Since these pieces are the same length, these two oppositely directed forces have the same magnitude, $I\ell_1 B = I\ell_3 B$, and they cancel. So the net magnetic force on the wire is the magnetic force on the middle piece. Since I points to the left and **B** is out of the page, the right-hand rule tells us the force is upward.

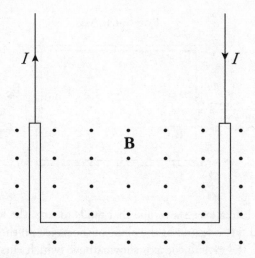

Since the magnetic force on the wire is $I\ell_2 B$, directed upward, the amount of current must create an upward magnetic force that exactly balances the downward gravitational force on the wire. Because the total mass of the wire is m, the resultant force (magnetic + gravitational) will be zero if

$$I\ell_2 B = mg \quad \Rightarrow \quad I = \frac{mg}{\ell_2 B}$$

Example 7 A rectangular loop of wire that carries a current I is placed in a uniform magnetic field, **B**, as shown in the diagram below and is free to rotate. What torque does it experience?

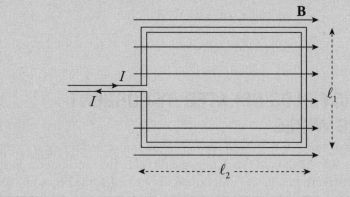

Solution. Ignoring the tiny gap in the vertical left-hand wire, we have two wires of length ℓ_1 and two of length ℓ_2. There is no magnetic force on either of the sides of the loop of length ℓ_2, because the current in the top side is parallel to **B** and the current in the bottom side is antiparallel to **B**. The magnetic force on the right-hand side points out of the plane of the page, while the magnetic force on the left-hand side points into the plane of the page.

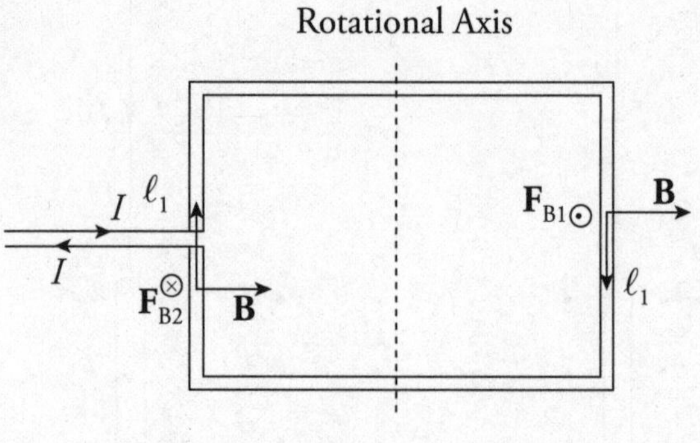

Rotational Axis

Each of these two forces exerts a torque that tends to turn the loop in such a way that the right-hand side rises out of the plane of the page and the left-hand side rotates into the page. Relative to the axis shown above (which cuts the loop in half), the torque of $\mathbf{F}_{B1}$ is

$$\tau_1 = r F_{B1} \sin \theta = \left(\frac{1}{2} \ell_2\right)(I \ell_1 B) \sin 90° = \frac{1}{2} I \ell_1 \ell_2 B$$

and the torque of $\mathbf{F}_{B2}$ is

$$\tau_2 = r F_{B2} \sin \theta = \left(\frac{1}{2} \ell_2\right)(I \ell_1 B) \sin 90° = \frac{1}{2} I \ell_1 \ell_2 B$$

Since both these torques rotate the loop in the same direction, the net torque on the loop is

$$\tau_1 + \tau_2 = I \ell_1 \ell_2 B$$

MAGNETIC FIELDS CREATED BY CURRENT-CARRYING WIRES

In the previous section, we examined the force that a current-carrying wire experiences when it is subjected to an external magnetic field. The source of the magnetic field in the previous section was unspecified. As we said at the beginning of this chapter, the sources of magnetic fields are electric charges that move; they may spin, circulate, move through space, or flow down a wire. For example, consider a long, straight wire that carries a current I. The current generates a magnetic field in the surrounding space, of magnitude

$$B = \frac{\mu_0}{2\pi} \frac{I}{r}$$

where r is the distance from the wire. The symbol μ_0 denotes a fundamental constant called the permeability of free space. Its value is:

$$\mu_0 = 4\pi \times 10^{-7}\,\text{N}\,/\,\text{A}^2 = 4\pi \times 10^{-7}\,\text{T} \cdot \text{m/A}$$

The magnetic field lines are actually circles whose centers are on the wire. The direction of these circles is determined by a variation of the right-hand rule. Imagine grabbing the wire in your right hand with your thumb pointing in the direction of the current. Then the direction in which your fingers curl around the wire gives the direction of the magnetic field line.

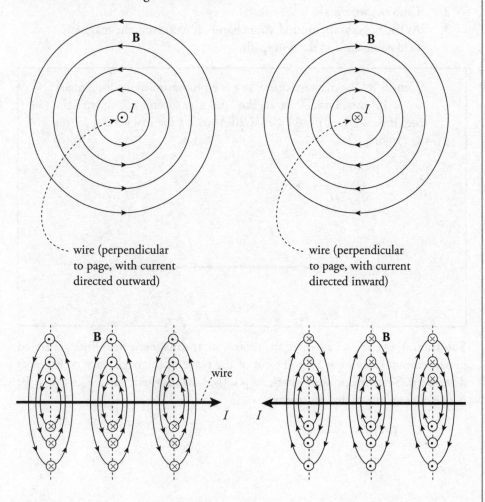

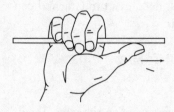 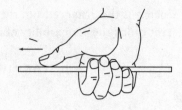

Right-Hand Rule for the Magnetic Field Created by a Current-Carrying Wire:

1. Put your thumb in direction of current or in direction of a positive traveling charge.
2. Grab the wire/path.
3. As the fingers curl around your thumb, it represents the magnetic field going around the wire/path.

Example 8 The diagram below shows a proton moving with a speed of 2×10^5 m/s, initially parallel to, and 4 cm from, a long, straight wire. If the current in the wire is 20 A, what's the magnetic force on the proton?

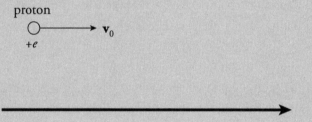

Solution. Above the wire (where the proton is), the magnetic field lines generated by the current-carrying wire point out of the plane of the page, so $\mathbf{v}_0 \times \mathbf{B}$ points downward. Since the proton's charge is positive, the magnetic force $\mathbf{F}_B = q(\mathbf{v}_0 \times \mathbf{B})$ is also directed down, toward the wire.

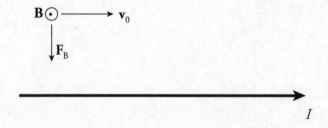

The strength of the magnetic force on the proton is

$$F_B = qv_0 B = ev_0 \frac{\mu_0}{2\pi} \frac{I}{r} = (1.6 \times 10^{-19} \text{ C})(2 \times 10^5 \text{ m/s}) \frac{4\pi \times 10^{-7} \text{ N/A}^2}{2\pi} \frac{20 \text{ A}}{0.04 \text{ m}}$$

$$= 3.2 \times 10^{-18} \text{ N}$$

Example 9 The diagram below shows a pair of long, straight, parallel wires, separated by a small distance, r. If currents I_1 and I_2 are established in the wires, what is the magnetic force per unit length they exert on each other?

Wire 1 ——————————————————→ I_1

Wire 2 ——————————————————→ I_2

Solution. To find the force on Wire 2, consider the current in Wire 1 as the source of the magnetic field. Below Wire 1, the magnetic field lines generated by Wire 1 point into the plane of the page. Therefore, the force on Wire 2, as given by the equation $\mathbf{F}_{B2} = I_2(\ell_2 \times \mathbf{B}_1)$, points upward.

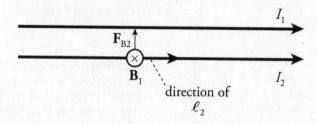

The magnitude of the magnetic force per unit length felt by Wire 2, due to the magnetic field generated by Wire 1, is found this way:

$$F_{B2} = I_2\ell_2 B_1 = I_2\ell_2 \frac{\mu_0}{2\pi}\frac{I_1}{r} \implies \frac{F_{B2}}{\ell_2} = \frac{\mu_0}{2\pi}\frac{I_1 I_2}{r}$$

By Newton's Third Law, this is the same force that Wire 1 feels due to the magnetic field generated by Wire 2. The force is attractive because the currents point in the same direction; if one of the currents were reversed, then the force between the wires would be repulsive.

Chapter 8 Review Questions

Solutions can be found in Chapter 12.

Section I: Multiple Choice

1. Which of the following is/are true concerning magnetic forces and fields? Select two answers.

   (A) The magnetic field lines due to a current-carrying wire radiate away from the wire.
   (B) The kinetic energy of a charged particle can be increased by a magnetic force.
   (C) A charged particle can move through a magnetic field without feeling a magnetic force.
   (D) A moving charged particle generates a magnetic field.

2. The velocity of a particle of charge $+4.0 \times 10^{-9}$ C and mass 2×10^{-4} kg is perpendicular to a 0.1-tesla magnetic field. If the particle's speed is 3×10^4 m/s, what is the acceleration of this particle due to the magnetic force?

   (A) 0.0006 m/s^2
   (B) 0.006 m/s^2
   (C) 0.06 m/s^2
   (D) 0.6 m/s^2

3. In the figure below, what is the direction of the magnetic force $\mathbf{F}_B$?

   (A) Downward, in the plane of the page
   (B) Upward, in the plane of the page
   (C) Out of the plane of the page
   (D) Into the plane of the page

4. In the figure below, what must be the direction of the particle's velocity, $\mathbf{v}$?

   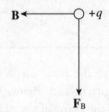

   (A) Downward, in the plane of the page
   (B) Upward, in the plane of the page
   (C) Out of the plane of the page
   (D) Into the plane of the page

5. Due to the magnetic force, a positively charged particle executes uniform circular motion within a uniform magnetic field, $\mathbf{B}$. If the charge is q and the radius of its path is r, which of the following expressions gives the magnitude of the particle's linear momentum?

   (A) qBr
   (B) qB/r
   (C) $q/(Br)$
   (D) $B/(qr)$

6. A straight wire of length 2 m carries a 10-amp current. How strong is the magnetic field at a distance of 2 cm from the wire?

   (A) 1×10^{-5} T
   (B) 2×10^{-5} T
   (C) 1×10^{-4} T
   (D) 2×10^{-4} T

7. Two long, straight wires are hanging parallel to each other and are 1 cm apart. The current in Wire 1 is 5 A, and the current in Wire 2 is 10 A, in the same direction. Which of the following best describes the magnetic force per unit length felt by the wires?

   (A) The force per unit length on Wire 1 is twice the force per unit length on Wire 2.
   (B) The force per unit length on Wire 2 is twice the force per unit length on Wire 1.
   (C) The force per unit length on Wire 1 is 0.0003 N/m, away from Wire 2.
   (D) The force per unit length on Wire 1 is 0.001 N/m, toward Wire 2.

8. In the figure below, what is the magnetic field at the Point P, which is midway between the two wires?

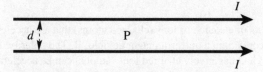

(A) $2\mu_0 I/(\pi d)$, into the plane of the page
(B) $\mu_0 I/(2\pi d)$, out of the plane of the page
(C) $\mu_0 I/(2\pi d)$, into the plane of the page
(D) Zero

9. Here is a section of a wire with a current moving to the right. Where is the magnetic field strongest and pointing INTO the page?

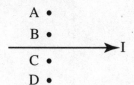

(A) A
(B) B
(C) C
(D) D

10. What is the direction of force acting on the current-carrying wire as shown below?

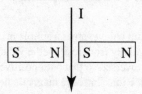

(A) To the bottom of the page
(B) Into the page
(C) Out of the page
(D) To the right of the page

Section II: Free Response

1. The diagram below shows a simple mass spectrograph. It consists of a source of ions (charged atoms) that are accelerated (essentially from rest) by the voltage *V* and enter a region containing a uniform magnetic field, **B**. The polarity of *V* may be reversed so that both positively charged ions (cations) and negatively charged ions (anions) can be accelerated. Once the ions enter the magnetic field, they follow a semicircular path and strike the front wall of the spectrograph, on which photographic plates are constructed to record the impact. Assume that the ions have mass *m*.

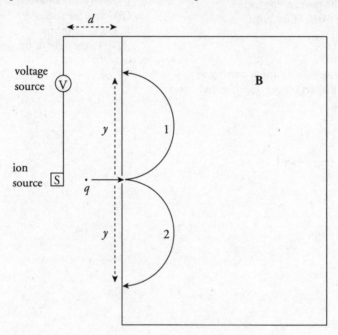

(a) What is the acceleration of an ion of charge *q* just before it enters the magnetic field?

(b) Find the speed with which an ion of charge *q* enters the magnetic field.

(c) (i) Which semicircular path, 1 or 2, would a cation follow?

    (ii) Which semicircular path, 1 or 2, would an anion follow?

(d) Determine the mass of a cation entering the apparatus in terms of *y*, *q*, **B**, and *V*.

(e) Once a cation of charge *q* enters the magnetic field, how long does it take to strike the photographic plate?

(f) What is the work done by the magnetic force in the spectrograph on a cation of charge *q*?

2. A particle accelerator has a collision that results in a photon, an anti-bottom quark, and a charm quark. The magnetic field is 6.00×10^{-8} T and can be described as into the page. A photon has no charge and has an upper theoretical mass of 3.6×10^{-52} kg. The charm quark has a mass of 2.23×10^{-27} kg, a charge of 1.07×10^{-19} C, and a velocity of 40.1 m/s. The anti-bottom quark has a mass of 7.49×10^{-27} kg and orbits with a radius of 92.7 m at a velocity of 41.5 m/s in a clockwise manner.

(a)   What is the orbital radius of the photon?

(b)   What is the orbital radius of the charm quark?

(c)   What is the charge of the anti-bottom quark?

Summary

o Charges moving though a magnetic field experience a force whose magnitude is given by $F_B = qv\mathbf{B}\sin\theta$ and whose direction is given by the right-hand rule.

o Because the force is always perpendicular to the direction of velocity, the charge may experience uniform circular motion. It would then follow all the appropriate circular motion relationships and orbit in a radius given by $r = \dfrac{mv}{q\mathbf{B}}$.

o Because wires have charges moving though them, a wire will experience a force if placed in a magnetic field. This is expressed by $F_B = \mathbf{B}I\ell\sin\theta$. Also, a current-carrying wire will produce a magnetic field whose strength is given by $\mathbf{B} = \dfrac{\mu_o}{2\pi}\dfrac{I}{r}$, where $\mu_0 = 4\pi \times 10^{-7}$ T·m/A, I is the current through the wire, and r is the radial distance from the wire.

Chapter 9
Electromagnetic Induction

INTRODUCTION

In Chapter 8, we learned that electric currents generate magnetic fields, and we will now see how magnetism can generate electric currents.

MOTIONAL EMF

Vice-Versa

In the previous chapter we learned moving charges generate magnetic fields. If we think about this backwards, magnetic fields then can generate moving charges or current.

The figure below shows a conducting wire of length ℓ, moving with constant velocity $\mathbf{v}$ in the plane of the page through a uniform magnetic field $\mathbf{B}$ that's perpendicular to the page. The magnetic field exerts a force on the moving conduction electrons in the wire. With $\mathbf{B}$ pointing into the page, the direction of $\mathbf{v}$ is to the right, so the magnetic force, $\mathbf{F}_B$, for positive charges, would be upward by the right-hand rule, so $\mathbf{F}_B$ on these electrons (which are negatively charged) is downward.

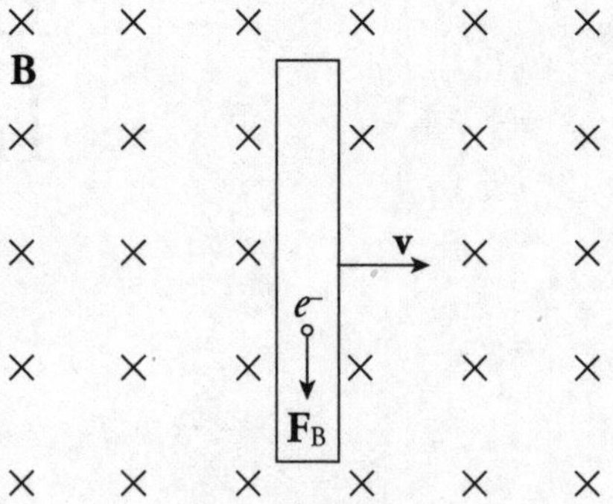

As a result, electrons will be pushed to the lower end of the wire, which will leave an excess of positive charge at its upper end. This separation of charge creates a uniform electric field, $\mathbf{E}$, within the wire (pointing downward).

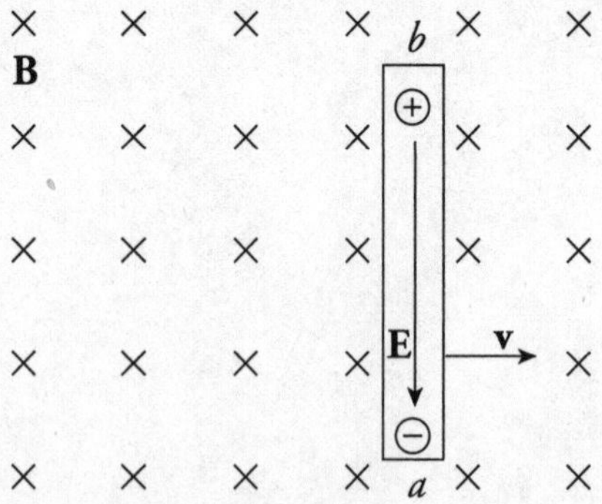

Right-Hand Rules

You will have to use a combination of your right-hand rules in order to find the direction of current. Just pay attention to orientation.

A charge q in the wire feels two forces: an electric force, $\mathbf{F}_E = q\mathbf{E}$, and a magnetic force, $F_B = qvB\sin\theta = qvB$, because $\theta = 90°$.

If q is negative, $\mathbf{F}_E$ is upward and $\mathbf{F}_B$ is downward; if q is positive, $\mathbf{F}_E$ is downward and $\mathbf{F}_B$ is upward. So, in both cases, the forces act in opposite directions. Once the magnitude of $\mathbf{F}_E$ equals the magnitude of $\mathbf{F}_B$, the charges in the wire are in electromagnetic equilibrium. This occurs when $qE = qvB$; that is, when $E = vB$.

The presence of the electric field creates a potential difference between the ends of the rod. Since negative charge accumulates at the lower end (which we'll call point *a*) and positive charge accumulates at the upper end (point *b*), point *b* is at a higher electric potential. The potential difference V_{ba} is equal to $E\ell$ and, since $E = vB$, the potential difference can be written as $vB\ell$.

Now, imagine that the rod is sliding along a pair of conducting rails connected at the left by a stationary bar. The sliding rod now completes a rectangular circuit, and the potential difference V_{ba} causes current to flow.

Induced Current

An induced current can be created three different ways:

1. Changing the area of the loop of wire in a stationary magnetic field
2. Changing the magnetic field strength through a stationary circuit
3. Changing the angle between the magnetic field and the wire loop

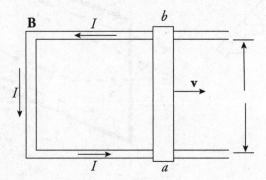

Each of these three changes causes a change in the flux, as we will see in the next section. The motion of the sliding rod through the magnetic field creates an electromotive force, called **motional emf**:

$$\mathcal{E} = B\ell v$$

Equation Sheet

The existence of a current in the sliding rod causes the magnetic field to exert a force on it. Using the formula $F_B = BI\ell$, the fact that ℓ points upward (in the direction of the current) and **B** is into the page, tells us that the direction of $\mathbf{F}_B$ on the rod is to the left. An external agent must provide this same amount of force to the right to maintain the rod's constant velocity and keep the current flowing. The power that the external agent must supply is $P = Fv = I\ell Bv$, and the electrical power delivered to the circuit is $P = IV_{ba} = I\mathcal{E} = IvB\ell$. Notice that these two expressions are identical. The energy provided by the external agent is transformed first into electrical energy and then thermal energy as the conductors making up the circuit dissipate heat.

This relationship between current in a coil of wire and magnetic fields sets the basis for Faraday's Law.

FARADAY'S LAW OF ELECTROMAGNETIC INDUCTION

Electromotive force can be created by the motion of a conducting wire through a magnetic field, but there is another way to generate an emf from a magnetic field.

Faraday's discovery found that a current is induced when the magnetic flux passing through the coil or loop of wire changes. Magnetic flux helps us define what that amount of field passing into the loop means.

Imagine holding a loop of wire in front of a fan as shown:

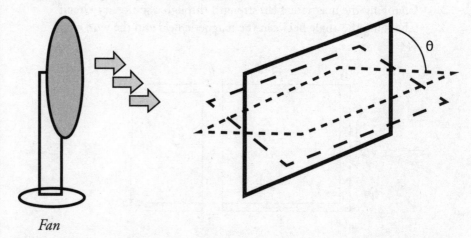

Fan

The amount of air that flows through the loop depends on the area of the loop as well as its orientation (tilt angle θ).

Loop seen from side

The most effective airflow is when the loop is completely perpendicular, as in the situation to the left. The least effective is when the airflow and loop are in the situation to the right.

We can apply this idea to a magnetic field passing through a loop. The magnetic flux, Φ_B, through an area A is equal to the product of A and the magnetic field parallel to the area vector. The area vector points normal to the surface.

$$\Phi_B = \mathbf{B} \bullet \mathbf{A} = BA \cos \theta$$

Equation Sheet

Magnetic flux measures the density of magnetic field lines that cross through an area. (Note that the direction of A is taken to be perpendicular to the plane of the loop.)

Magnetic Flux Units
The SI unit for Magnetic Flux is called the weber (Wb) which is equivalent to one Tesla meter-squared (Tm²).

Example 1 The figure below shows two views of a circular loop of radius 3 cm placed within a uniform magnetic field, **B** (magnitude 0.2 T).

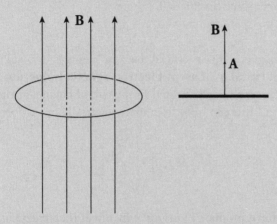

(a) What's the magnetic flux through the loop?
(b) What would be the magnetic flux through the loop if the loop were rotated 45°?
(c) What would be the magnetic flux through the loop if the loop were rotated 90°?

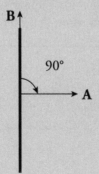

Solution.

(a) Since **B** is parallel to **A**, the magnetic flux is equal to BA:

$$\Phi_B = BA = B \cdot \pi r^2 = (0.2 \text{ T}) \cdot \pi(0.03 \text{ m})^2 = 5.7 \times 10^{-4} \text{ T} \cdot \text{m}^2$$

The SI unit for magnetic flux, the tesla·meter^2, is called a **weber** (abbreviated **Wb**). So $\Phi_B = 5.7 \times 10^{-4}$ Wb.

(b) Since the angle between **B** and **A** is 45°, the magnetic flux through the loop is

$$\Phi_B = BA \cos 45° = B \cdot \pi r^2 \cos 45° = (0.2 \text{ T}) \cdot \pi(0.03 \text{ m})^2 \cos 45° = 4.0 \times 10^{-4} \text{ Wb}$$

(c) If the angle between **B** and **A** is 90°, the magnetic flux through the loop is zero, since $\cos 90° = 0$.

The concept of magnetic flux is crucial, because changes in magnetic flux induce emf. According to **Faraday's Law of Electromagnetic Induction**, the magnitude of the emf induced in a circuit is equal to the rate of change of the magnetic flux through the circuit. This can be written mathematically in the form

$$\left| \varepsilon_{\text{avg}} \right| = \left| \frac{\Delta \Phi_B}{\Delta t} \right|$$

This induced emf can produce a current, which will then create its own magnetic field. The direction of the induced current is determined by the polarity of the induced emf and is given by **Lenz's Law**: The induced current will always flow in the direction that opposes the change in magnetic flux that produced it. If this were not so, then the magnetic flux created by the induced current would magnify the change that produced it, and energy would not be conserved. Lenz's Law can be included mathematically with Faraday's Law by the introduction of a minus sign; this leads to a single equation that expresses both results:

Equation Sheet

$$\varepsilon_{\text{avg}} = -\frac{\Delta \Phi_B}{\Delta t}$$

Example 2 The circular loop of Example 1 rotates at a constant angular speed through 45° in 0.5 s.

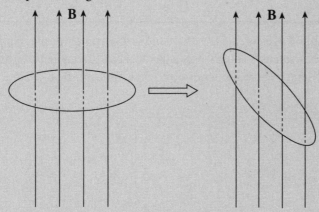

(a) What's the induced emf in the loop?

(b) In which direction will current be induced to flow?

Solution.

(a) As we found in Example 1, the magnetic flux through the loop changes when the loop rotates. Using the values we determined earlier, Faraday's Law gives

$$\mathcal{E}_{avg} = -\frac{\Phi_B}{t} = -\frac{(4.0 \times 10^{-4} \text{ Wb}) - (5.7 \times 10^{-4} \text{ Wb})}{0.5 \text{ s}} = 3.4 \times 10^{-4} \text{ V}$$

(b) The original magnetic flux was 5.7×10^{-4} Wb upward, and was decreased to 4.0×10^{-4} Wb. So the change in magnetic flux is -1.7×10^{-4} Wb upward, or, equivalently, $\Delta\Phi_B = 1.7 \times 10^{-4}$ Wb, downward. To oppose this change we would need to create some magnetic flux upward. The current would be induced in the counter-clockwise direction (looking down on the loop), because the right-hand rule tells us that then the current would produce a magnetic field that would point up.

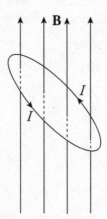

The current will flow only while the loop rotates, because emf is induced only when magnetic flux is changing. If the loop rotates 45° and then stops, the current will disappear.

Example 3 Again consider the conducting rod that's moving with constant velocity **v** along a pair of parallel conducting rails (separated by a distance ℓ), within a uniform magnetic field, **B**:

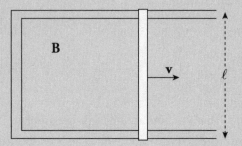

Find the induced emf and the direction of the induced current in the rectangular circuit.

Solution. The area of the rectangular loop is ℓx, where x is the distance from the left-hand bar to the moving rod:

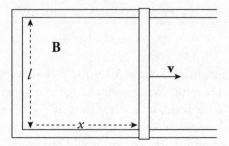

Because the area is changing, the magnetic flux through the loop is changing, which means that an emf will be induced in the loop. To calculate the induced emf, we first write $\Phi_{B} = BA = B \ell x$, then since $\Delta x/\Delta t = v$, we get

$$\varepsilon_{\text{avg}} = -\frac{\Delta \Phi_{B}}{\Delta t} = -\frac{\Delta(B\ell x)}{\Delta t} = -B\ell \frac{\Delta x}{\Delta t} = -B\ell v$$

We can figure out the direction of the induced current from Lenz's Law. As the rod slides to the right, the magnetic flux into the page increases. How do we oppose an increasing into-the-page flux? By producing out-of-the-page flux. In order for the induced current to generate a magnetic field that points out of the plane of the page, the current must be directed counterclockwise (according to the right-hand rule).

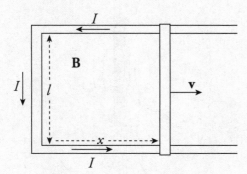

Note that the magnitude of the induced emf and the direction of the current agree with the results we derived earlier, in the section on motional emf.

This example also shows how a violation of Lenz's Law would lead directly to a violation of the Law of Conservation of Energy. The current in the sliding rod is directed upward, as given by Lenz's Law, so the conduction electrons are drifting downward. The force on these drifting electrons—and thus, the rod itself—is directed to the left, opposing the force that's pulling the rod to the right. If the current were directed downward, in violation of Lenz's Law, then the magnetic force on the rod would be to the right, causing the rod to accelerate to the right with ever-increasing speed and kinetic energy, without the input of an equal amount of energy from an external agent.

Example 4 A permanent magnet creates a magnetic field in the surrounding space. The end of the magnet at which the field lines emerge is designated the **north pole** (N), and the other end is the **south pole** (S):

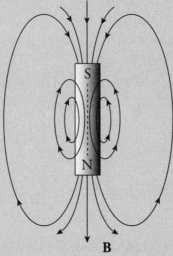

B

(a) The figure below shows a bar magnet moving down, through a circular loop of wire. What will be the direction of the induced current in the wire?

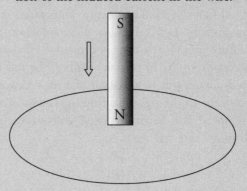

(b) What will be the direction of the induced current in the wire if the magnet is moved as shown in the following diagram?

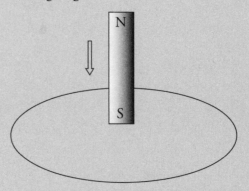

Solution.

(a)   The magnetic flux down, through the loop, increases as the magnet is moved. By Lenz's Law, the induced emf will generate a current that opposes this change. How do we oppose a change of *more flux downward*? By creating flux *upward*. So, according to the right-hand rule, the induced current must flow counterclockwise (because this current will generate an upward-pointing magnetic field):

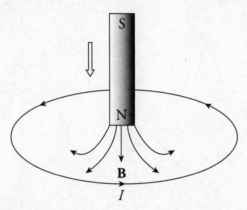

(b)   In this case, the magnetic flux through the loop is upward and, as the south pole moves closer to the loop, the magnetic field strength increases so the magnetic flux through the loop increases upward. How do we oppose a change of *more flux upward*? By creating flux *downward*. Therefore, in accordance with the right-hand rule, the induced current will flow clockwise (because this current will generate a downward-pointing magnetic field):

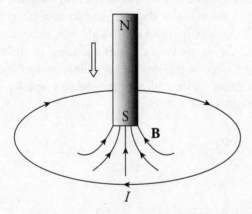

Example 5 A square loop of wire 2 cm on each side contains 5 tight turns and has a total resistance of 0.0002 Ω. It is placed 20 cm from a long, straight, current-carrying wire. If the current in the straight wire is increased at a steady rate from 20 A to 50 A in 2 s, determine the magnitude and direction of the current induced in the square loop. (Because the square loop is at such a great distance from the straight wire, assume that the magnetic field through the loop is uniform and equal to the magnetic field at its center.)

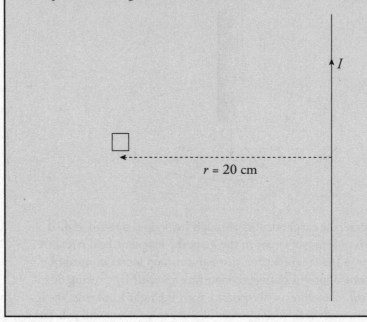

$r = 20$ cm

Solution. At the position of the square loop, the magnetic field due to the straight wire is directed out of the plane of the page and its strength is given by the equation $B = (\mu_0/2\pi)(I/r)$. As the current in the straight wire increases, the magnetic flux through the turns of the square loop changes, inducing an emf and current. There are N = 5 turns; each loop contributes the same flux, so the total flux becomes the number of loops N times the flux of each individual loop, Φ_B. Faraday's Law becomes $\mathcal{E}_{avg} = -N(\Delta\Phi_B/\Delta t)$, and

$$\mathcal{E}_{avg} = -N\frac{\Delta\Phi_B}{\Delta t} = -N\frac{\Delta(BA)}{\Delta t} = -NA\frac{\Delta B}{\Delta t} = -NA\frac{\mu_0}{2\pi r}\frac{\Delta I}{\Delta t}$$

Substituting the given numerical values, we get

$$\mathcal{E}_{avg} = -NA\frac{\mu_0}{2\pi r}\frac{\Delta I}{\Delta t}$$
$$= -(5)(0.02 \text{ m})^2 \frac{4\pi \times 10^{-7} \text{ T} \cdot \text{m/A}}{2\pi(0.20 \text{ m})} \frac{(50 \text{ A} - 20 \text{ A})}{2 \text{ s}}$$
$$= -3 \times 10^{-8} \text{ V}$$

The magnetic flux through the loop is out of the page and increases as the current in the straight wire increases. To oppose an increasing out-of-the-page flux, the direction of the induced current should be clockwise, thereby generating an into-the-page magnetic field (and flux).

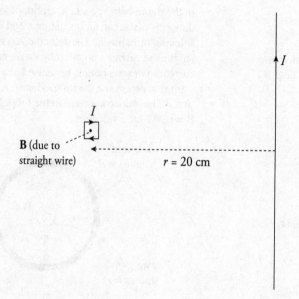

The value of the current in the loop will be

$$I = \frac{\mathcal{E}}{R} = \frac{3 \times 10^{-8} \text{ V}}{0.0002} = 1.5 \times 10^{-4} \text{ A}$$

Chapter 9 Review Questions

Solutions can be found in Chapter 12.

Section I: Multiple Choice

1. A metal rod of length L is pulled upward with constant velocity v through a uniform magnetic field $\mathbf{B}$ that points out of the plane of the page.

B

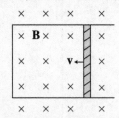

What is the potential difference between points a and b ?

(A) 0

(B) $\dfrac{1}{2}vBL$, with point b at the higher potential

(C) vBL, with point a at the higher potential

(D) vBL, with point b at the higher potential

2. A conducting rod of length 0.2 m and resistance 10 ohms between its endpoints slides without friction along a U-shaped conductor in a uniform magnetic field B of magnitude 0.5 T perpendicular to the plane of the conductor, as shown in the diagram below.

If the rod is moving with velocity $\mathbf{v}$ = 3 m/s to the left, what is the magnitude and direction of the current induced in the rod?

	Current	Direction
(A)	0.03 A	down
(B)	0.03 A	up
(C)	0.3 A	down
(D)	0.3 A	up

3. In the figure below, a small, circular loop of wire (radius r) is placed on an insulating stand inside a hollow solenoid of radius R. The solenoid has n turns per unit length and carries a counterclockwise current I. If the current in the solenoid is decreased at a steady rate of a amps/s, determine the induced emf, ε, and the direction of the induced current in the loop. Note that $B = \mu_0\, n\, I$ for a solenoid.

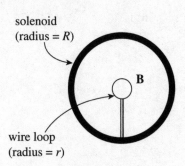

solenoid (radius = R)

B

wire loop (radius = r)

(A) $\varepsilon = \mu_0 \pi n r^2 a$; induced current is clockwise

(B) $\varepsilon = \mu_0 \pi n r^2 a$; induced current is counterclockwise

(C) $\varepsilon = \mu_0 \pi n R^2 a$; induced current is clockwise

(D) $\varepsilon = \mu_0 \pi n R^2 a$; induced current is counterclockwise

4. In the figures below, a permanent bar magnet is below a loop of wire. It is pulled upward with a constant velocity through the loop of wire as shown in Figure b.

Figure a

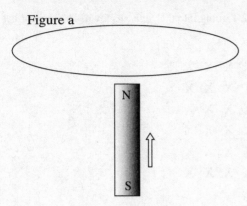

Figure b

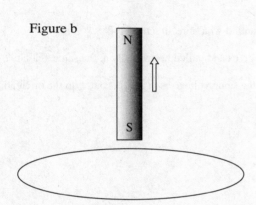

Which of the following best describes the direction(s) of the current induced in the loop (looking down on the loop from above)?

(A) Always clockwise
(B) Always counterclockwise
(C) First clockwise, then counterclockwise
(D) First counterclockwise, then clockwise

5. A square loop of wire (side length = s) surrounds a long, straight wire such that the wire passes through the center of the square.

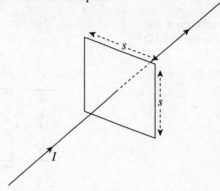

If the current in the wire is I, determine the current induced in the square loop.

(A) $\dfrac{2\mu_0 Is}{\pi\left(1+\sqrt{2}\right)}$

(B) $\dfrac{\mu_0 Is}{\pi\sqrt{2}}$

(C) $\dfrac{\mu_0 Is\sqrt{2}}{\pi}$

(D) 0

Section II: Free Response

1. A rectangular wire is pulled through a uniform magnetic field of 2 T going into the page as shown. The resistor has a resistance of 20 Ω.

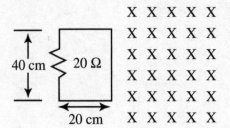

(a)   What is the voltage across the resistor as the wire is pulled horizontally at a velocity of 1 m/s and it just enters the field?

(b)   What is the current through the circuit in the above case and in what direction does it flow?

(c)   What would the voltage be if the wire were rotated 90 degrees and pulled horizontally at the same velocity?

(d)   What would the velocity have to be in order to maintain the same voltage as in part (a) but with the orientation of part (c)?

Summary

o An electromotive force is produced as a conducting wire's position changes or as the magnetic field changes. This idea is summarized by $\varepsilon = B\ell v$.

o The flux (ϕ) tells the amount of something (for this chapter the magnetic field) that goes through a surface. The magnetic flux depends on the strength of the magnetic field, the surface area through which the field passes, and the angle between the two. This idea is summed up by $\phi_B = BA\cos\theta$.

o Faraday's Law of Induction says that if the wire is formed in a loop, an electromagnetic force is produced if the magnetic flux changes with time. This idea can be summarized by $\varepsilon_{avg} = -\dfrac{\Delta\phi_B}{\Delta t}$.

126

Chapter 10
Optics

INTRODUCTION

Light (or visible light) makes up only a small part of the entire spectrum of electromagnetic waves, which ranges from radio waves to gamma rays. Most waves require a material medium for their transmission, but electromagnetic waves can propagate through empty space. Electromagnetic waves consist of time-varying electric and magnetic fields that oscillate perpendicular to each other and to the direction of propagation of the wave. Through a vacuum, all electromagnetic waves travel at a fixed speed:

$$c = 3.00 \times 10^8 \text{ m/s}$$

regardless of their frequency. Like all the waves, we can say $v = f\lambda$.

ELECTROMAGNETIC WAVES

Back in AP Physics 1, we saw that if we oscillate one end of a long rope, we generate a wave that travels down the rope and has the same frequency as that of the oscillation.

You can think of an electromagnetic wave in a similar way: An oscillating electric charge generates an electromagnetic (EM) wave, which is composed of oscillating electric and magnetic fields. These fields oscillate with the same frequency at which the electric charge that created the wave oscillated. The fields oscillate in phase with each other, perpendicular to each other and the direction of propagation. For this reason, electromagnetic waves are transverse waves. The direction in which the wave's electric field oscillates is called the direction of polarization of the wave.

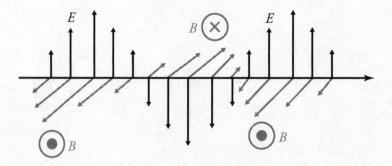

Unlike waves on a rope or sound waves, electromagnetic waves do not require a material medium to propagate; they can travel through empty space (vacuum). When an EM wave travels through a vacuum, its speed is constant, $c = 3 \times 10^8$ m/s.

EM waves obey the classic wave equation:

$$v = \lambda f$$

Equation Sheet

THE ELECTROMAGNETIC SPECTRUM

Electromagnetic waves can be categorized by their frequency (or wavelength); the full range of waves is called the **electromagnetic** (or **EM**) **spectrum**. Types of waves include **radiowaves**, **microwaves**, **infrared**, **visible light**, **ultraviolet**, **X-rays**, and **γ-rays** (**gamma rays**) and, although they've been delineated in the spectrum below, there's no universal agreement on all the boundaries, so many of these bands overlap. You should be familiar with the names of the major categories, and, in particular, memorize the order of the colors within the visible spectrum (which, as you can see, accounts for only a tiny sliver of the full EM spectrum). In order of increasing wave frequency, the colors are red, orange, yellow, green, blue, and violet, which is commonly remembered as ROYGBV ("roy-gee-biv"). The wavelengths of the colors in the visible spectrum are usually expressed in nanometers. For example, electromagnetic waves whose wavelengths are between 577 nm and 597 nm are seen as yellow light.

Electromagnetic Wave Speed
All electromagnetic waves, regardless of frequency, travel through a vacuum at this speed. The most important equation for waves $v = \lambda f$, is also true for electromagnetic waves. For EM waves traveling through a vacuum, $v = c$, so the equation becomes $\lambda f = c$.

ELECTROMAGNETIC SPECTRUM

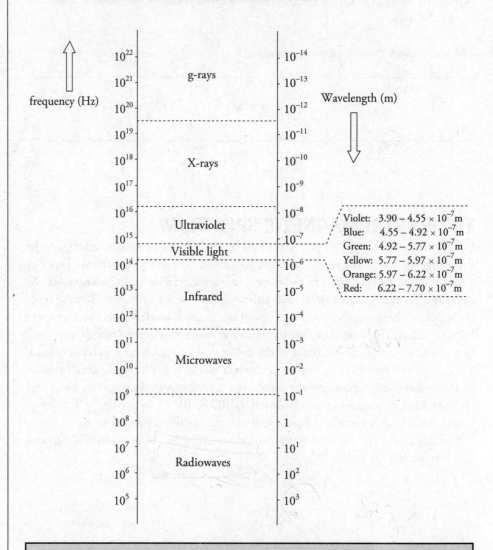

Violet:	$3.90 - 4.55 \times 10^{-7}$ m
Blue:	$4.55 - 4.92 \times 10^{-7}$ m
Green:	$4.92 - 5.77 \times 10^{-7}$ m
Yellow:	$5.77 - 5.97 \times 10^{-7}$ m
Orange:	$5.97 - 6.22 \times 10^{-7}$ m
Red:	$6.22 - 7.70 \times 10^{-7}$ m

Example 1 What's the frequency range for green light?

Solution. According to the spectrum, light is green if its wavelength is between 4.92×10^{-7} m and 5.77×10^{-7} m. Using the equation $v = f\lambda$, we find that the upper end of this wavelength range corresponds to a frequency of

$$f_1 = \frac{v}{\lambda_1} = \frac{3.00 \times 10^8 \text{ m/s}}{5.77 \times 10^{-7} \text{ m}} = 5.20 \times 10^{14} \text{ Hz}$$

while the lower end corresponds to

$$f_2 = \frac{v}{\lambda_2} = \frac{3.00 \times 10^8 \text{ m/s}}{4.92 \times 10^{-7} \text{ m}} = 6.10 \times 10^{14} \text{ Hz}$$

Equation Sheet

So the frequency range for green light is

$$5.20 \times 10^{14} \text{ Hz} \leq f_{\text{green}} \leq 6.10 \times 10^{14} \text{ Hz}$$

> **Example 2** How would you classify electromagnetic radiation that has a wavelength of 1 cm?

Solution. According to the electromagnetic spectrum presented above, electromagnetic waves with $\lambda = 10^{-2}$ m are microwaves.

INTERFERENCE AND DIFFRACTION

As we learned in AP Physics 1, waves experience interference when they meet, and whether they interfere constructively or destructively depends on their relative phase. If they meet *in phase* (crest meets crest), they combine constructively, but if they meet *out of phase* (crest meets trough), they combine destructively. The key to the interference patterns we'll study in the next section rests on this observation. In particular, if waves that have the same wavelength meet, then the difference in the distances they've traveled determine whether the waves are in phase. Assuming that the waves are **coherent** (which means that their phase difference remains constant over time and does not vary), if the difference in their path lengths, $\Delta \ell$, is a whole number of wavelengths—0, ±λ, ±2λ, etc.—they'll arrive in phase at the meeting point. On the other hand, if this difference is a whole number plus one-half a wavelength—±$\frac{1}{2}\lambda$, ±$(1 + \frac{1}{2})\lambda$, ±$(2 + \frac{1}{2})\lambda$, etc.—then they'll arrive exactly out of phase. That is,

constructive interference: $\ell = m\lambda$

destructive interference: $\ell = (m + \frac{1}{2})\lambda$ where $m = 0, 1, 2\ldots$

Young's Double-Slit Interference Experiment

The following figure shows light incident on a barrier that contains two narrow slits (perpendicular to the plane of the page), separated by a distance d. On the right is a screen whose distance from the barrier, L, is much greater than d. The question is, what will we see on the screen? You might expect that we'll just see two bright narrow strips of light, directly opposite the slits in the barrier. As reasonable as this may sound, it doesn't take into account the wave nature of light.

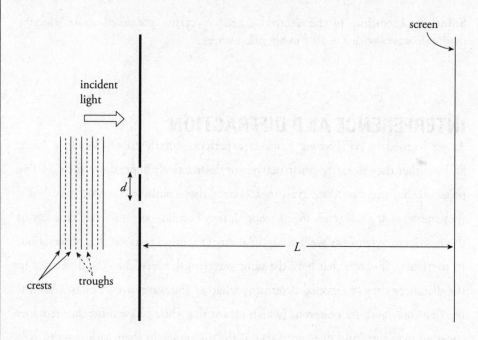

When a wave encounters an aperture whose width is comparable to its wavelength, the wave will fan out after it passes through. The alteration in the straight-line propagation of a wave when it encounters a barrier is called **diffraction**. In the set-up above, the waves will diffract through the slits, and spread out and interfere as they travel toward the screen.

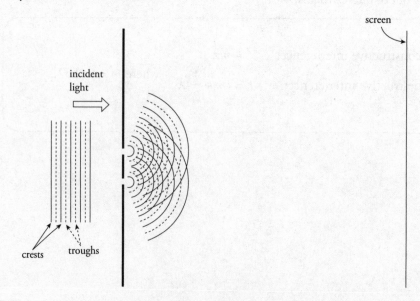

You can clearly see points of interference in the figure above. Look for solid lines intersecting other solid lines—these are points of constructive interference. Look for solid lines intersecting dashed lines—these are points of destructive interference.

The screen will show the results of this interference: There will be bright bands (bright **fringes**) centered at those points at which the waves interfere constructively, alternating with dark fringes, where the waves interfere destructively. Let's determine the locations of these fringes.

In the figure below, we've labeled the slits S_1 and S_2. A point P on the screen is selected, the path lengths from S_1 and S_2 to P are called ℓ_1 and ℓ_2, respectively, and the angle that the line from the midpoint of the slits to P makes with the horizontal is θ. Segment S_1Q is perpendicular to line S_2P. Because L is so much larger than d, the angle that line S_2P makes with the horizontal is also approximately θ, which tells us that $\angle S_2S_1Q$ is approximately θ and that the path difference, $\Delta\ell = \ell_2 - \ell_1$, is nearly equal to S_2Q.

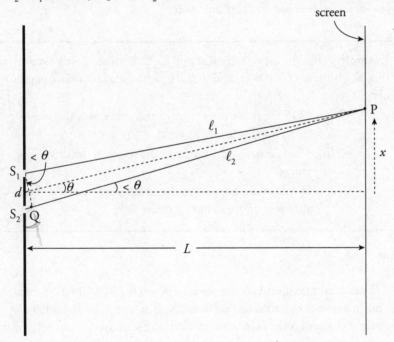

Because $S_2Q = d\sin\theta$, we get $\Delta\ell = d\sin\theta$. Now, using what we learned earlier about how constructive or destructive interference depends on $\Delta\ell$, we can write:

constructive interference
(intensity maximum
bright fringe on screen):
$$d\sin\theta = m\lambda$$
where $m = 0, 1, 2\ldots$

destructive interference
(intensity minimum
dark fringe on screen):
$$d\sin\theta = (m + \tfrac{1}{2}\lambda)$$

Equation Sheet

To locate the positions of, say, the bright fringes on the screen, we use the fact that $x = L \tan \theta$. If θ is small, then $\tan \theta \approx \sin \theta$, so we can write $x = L \sin \theta$ (we can tell this from the figure). Since $\sin \theta = m\lambda/d$ for bright fringes, we get

$$x_m = \frac{m\lambda L}{d}$$

Also, the intensity of the bright fringes decreases as m increases in magnitude. The bright fringe directly opposite the midpoint of the slits—the **central maximum**—will have the greatest intensity when $m = 0$. The bright fringes with $m = 1$ will have a lower intensity, those with $m = 2$ will be fainter still, and so on. If more than two slits are cut in the barrier, the interference pattern becomes sharper, and the distinction between dark and bright fringes becomes more pronounced. Barriers containing thousands of tiny slits per centimeter—called **diffraction gratings**—are used precisely for this purpose.

Example 3 For the experimental set-up we've been studying, assume that $d = 1.5$ mm, $L = 6.0$ m, and that the light used has a wavelength of 589 nm.

(a) How far above the center of the screen will the second brightest maximum appear?

(b) How far below the center of the screen is the third dark fringe?

(c) What would happen to the interference pattern if the slits were moved closer together?

Solution.

(a)   The central maximum corresponds to $m = 0$ ($x_0 = 0$). The first maximum above the central one is labeled x_1 (since $m = 1$). The other bright fringes on the screen are labeled accordingly.

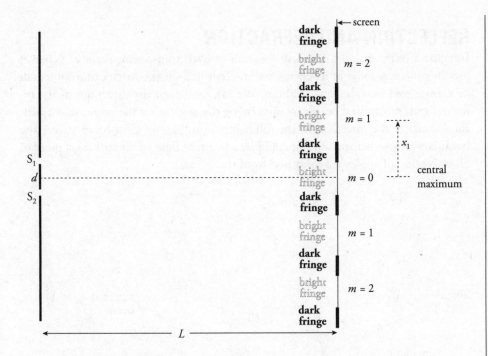

The value of x_1 is

$$x_1 = \frac{1 \cdot \lambda L}{d} = \frac{(589 \times 10^{-9} \text{ m})(6.0 \text{ m})}{1.5 \times 10^{-3} \text{ m}} = 2.4 \times 10^{-3} \text{ m} = 2.4 \text{ mm}$$

(b)   The first dark fringe occurs when the path difference is 0.5λ, the second dark fringe occurs when the path difference is 1.5λ, and the third dark fringe occurs when the path difference is 2.5λ, so

$$x_{\substack{\text{3rd minimum} \\ \text{below central max}}} = \frac{(2 + \frac{1}{2})\lambda L}{d} = \frac{(2 + \frac{1}{2})(589 \times 10^{-9} \text{ m})(6.0 \text{ m})}{1.5 \times 10^{-3} \text{ m}} = -5.9 \times 10^{-3} \text{ m} = -5.9 \text{ mm}$$

(c)   Since $x_m = m\lambda L/d$, a decrease in d would cause an increase in x_m. That is, the fringes would become larger; the interference pattern would be more spread out.

Single-Slit Diffraction

A diffraction pattern will also form on the screen if the barrier contains only one slit. The central maximum will be very pronounced, but lower-intensity maxima will also be seen because of interference from waves arriving from different locations within the slit itself. The width of the central maximum will become wider as the width of the slit is decreased.

For a circular pinhole aperture, the diffraction pattern will consist of a central, bright circular disk surrounded by rings of decreasing intensity.

REFLECTION AND REFRACTION

Imagine a beam of light directed toward a smooth transparent surface. When it hits this surface, some of its energy will be reflected off the surface and some will be transmitted into the new medium. We can figure out the directions of the reflected and transmitted beams by calculating the angles that the beams make with the normal to the interface. In the following figure, an incident beam strikes the boundary of another medium; it could be a beam of light in air striking a piece of glass. Notice all angles are measured from the normal.

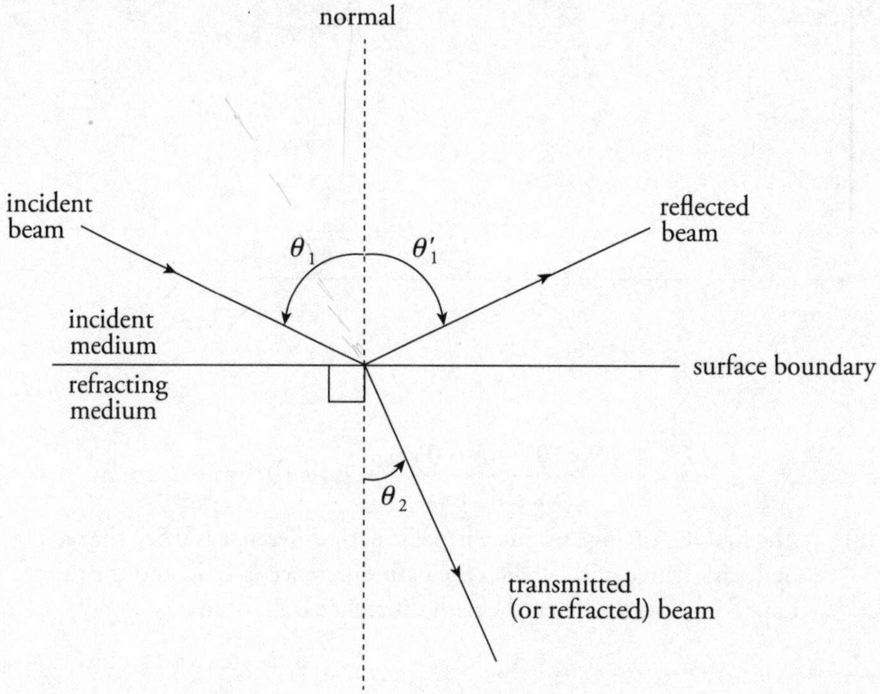

Tricky Problems!

Remember that the normal is always perpendicular to the surface. When a question asks for the angle of incidence or reflection, it wants the angle formed with the normal, not the angle to the horizontal surface. Make sure you know if you're solving for the angle of incidence, angle of reflection, or the angle it forms with the mirror.

The angle that the **incident beam** makes with the normal is called the **angle of incidence**, or θ_1. The angle that the **reflected beam** makes with the normal is called the **angle of reflection**, θ'_1, and the angle that the **transmitted beam** makes with the normal is called the **angle of refraction**, θ_2. The incident, reflected, and transmitted beams of light all lie in the same plane.

The relationship between θ_1 and θ'_1 is pretty easy; it is called the **Law of Reflection**:

$$\theta_1 = \theta'_1$$

The Law of Reflection basically states the Angle of Reflection is equal to the Angle of Incidence. In order to describe how θ_1 and θ_2 are related, we first need to talk about a medium's index of refraction.

When light travels through empty space (vacuum), its speed is $c = 3.00 \times 10^8$ m/s; this is one of the fundamental constants of nature. But when light travels through a material medium (such as water or glass), it's constantly being absorbed and re-emitted by the atoms that compose the material and, as a result, its apparent speed, v, is some fraction of c. The reciprocal of this fraction,

$$n = \frac{c}{v}$$

Equation Sheet

is called the medium's **index of refraction**. For example, since the speed of light in water is $v = 2.25 \times 10^8$ m/s, the index of refraction of water is

$$n = \frac{3.00 \times 10^8 \text{ m/s}}{2.25 \times 10^8 \text{ m/s}} = 1.33$$

Note that n has no units; it's also never less than 1, because light always travels slower in a medium than in a vacuum.

n of air
The index of refraction of air is 1.00029, which is so close to 1 that the norm is to consider it to just be 1.

The equation that relates θ_1 and θ_2 involves the index of refraction of the incident medium (n_1) and the index of refraction of the refracting medium (n_2); it's called **Snell's Law**:

$$n_1 \sin \theta_1 = n_2 \sin \theta_2$$

Equation Sheet

If $n_2 > n_1$, then Snell's Law tells us that $\theta_2 < \theta_1$; that is, the beam will bend (**refract**) *toward* the normal as it enters the medium. On the other hand, if $n_2 < n_1$, then $\theta_2 > \theta_1$, and the beam will bend *away* from the normal.

Example 4 A beam of light in air is incident upon a piece of glass, striking the surface at an angle of 30°. If the index of refraction of the glass is 1.5, what are the angles of reflection and refraction?

Solution. If the light beam makes an angle of 30° with the surface, then it makes an angle of 60° with the normal; this is the angle of incidence. By the Law of Reflection, then, the angle of reflection is also 60°. We use Snell's Law to find the angle of refraction. The index of refraction of air is close to 1, so we can say that $n = 1$ for air.

$$n_1 \sin \theta_1 = n_2 \sin \theta_2$$
$$(1) \sin 60° = 1.5 \sin \theta_2$$
$$\sin \theta_2 = 0.5774$$
$$\theta_2 = 35°$$

Note that $\theta_2 < \theta_1$, as we would expect since the refracting medium (glass) has a greater index than does the incident medium (air).

> **Example 5** A fisherman drops a flashlight into a lake that's 10 m deep. The flashlight sinks to the bottom where its emerging light beam is directed almost vertically upward toward the surface of the lake, making a small angle (θ_1) with the normal. How deep will the flashlight appear to be to the fisherman? (Use the fact that tan θ is almost equal to sin θ if θ is small.)

Solution. Take a look at the figure below.

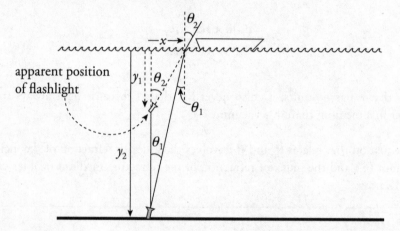

Since the refracting medium (air) has a lower index than the incident medium (water), the beam of light will bend away from the normal as it emerges from the water. As a result, the fisherman will think that the flashlight is at a depth of only y_1, rather than its actual depth of $y_2 = 10$ m. By simple trigonometry, we know that

$$\tan \theta_1 = \frac{x}{y_2} \qquad \text{and} \qquad \tan \theta_2 = \frac{x}{y_1}$$

So,

$$\frac{y_1}{y_2} = \frac{\tan \theta_1}{\tan \theta_2}$$

Snell's Law tells us that $n_1 \sin \theta_1 = \sin \theta_2$ (since $n_2 = n_{\text{air}} = 1$), so

$$\frac{\sin \theta_1}{\sin \theta_2} = \frac{1}{n_1}$$

Using the approximations $\sin \theta_1 \approx \tan \theta_1$ and $\sin \theta_2 \approx \tan \theta_2$, we can write

$$\frac{\tan \theta_1}{\tan \theta_2} \approx \frac{\sin \theta_1}{\sin \theta_2}$$

which means

$$\frac{y_1}{y_2} = \frac{1}{n_1}$$

Because $y_2 = 10$ m and $n_1 = n_{water} = 1.33$,

$$y_1 = \frac{y_2}{n_1} = \frac{10 \text{ m}}{1.33} = 7.5 \text{ m}$$

DISPERSION OF LIGHT

One thing we learned when we studied waves is that wave speed is independent of frequency. For a given medium, different frequencies give rise to different wavelengths, because the equation $v = f\lambda$ must always be satisfied and v doesn't vary. But when light travels through a material medium, it displays **dispersion**, which is a variation in wave speed with frequency (or wavelength). So, the definition of the index of refraction, $n = c/v$, should be accompanied by a statement of the frequency of the light used to measure v, since different frequencies have different speeds and different indices. A piece of glass may have the following indices for visible light:

	red	orange	yellow	green	blue	violet
$n =$	1.502	1.506	1.511	1.517	1.523	1.530

Note that as the wavelength decreases, refractive index increases. In general, higher frequency waves have higher indices of refraction. Most lists of refractive index values are tabulated using yellow light of wavelength 589 nm (frequency 5.1×10^{14} Hz).

Although the variation in the values of the refractive index across the visible spectrum is pretty small, when white light (which is a combination of all the colors of the visible spectrum) hits a glass prism, the beam is split into its component colors.

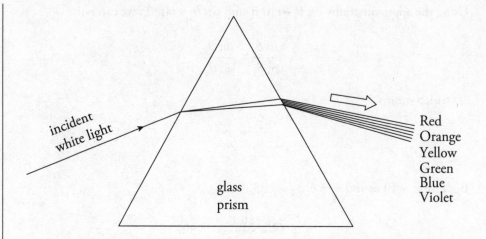

Why? Because each color has its own index, Snell's Law tells us that each color will have its own angle of refraction. Therefore, each color emerges from the prism at a slightly different angle, so the light disperses into its component colors.

> **Example 6** In the figure above, assume that the incident beam strikes the prism with an angle of incidence of 60.00°. If the top angle of the prism has a measure of 60.00°, what are the angles of refraction of red light and violet light when the beam emerges from the other side? (Use $n = 1.502$ for red light and 1.530 for violet light.)

Solution. The beam refracts twice, once when it enters the prism and again when it exits.

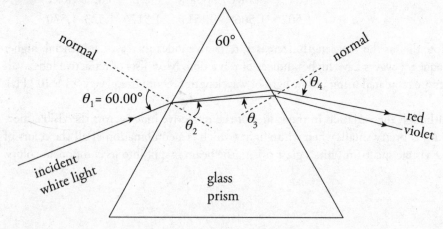

Snell's Law tells us that

$$\sin 60.00° = n \sin \theta_2 \text{ and } n \sin \theta_3 = \sin \theta_4$$

where n is the index of the glass. We'll ignore the minuscule variation in n for light in air, and just use $n_{air} = 1$ for all colors. The relationship between θ_2 and θ_3 follows from the fact that the sum of the measures of the angles in any triangle is 180°, so

$$(90° - \theta_2) + 60° + (90° - \theta_3) = 180°$$

which implies that $\theta_3 = 60° - \theta_2$. We can now make the following table of values:

color	n	θ_2	θ_3	θ_4
red	1.502	35.210°	24.790°	39.03°
violet	1.530	34.474°	25.526°	41.25°

In general, the differences in n are so small that, unless you are specifically addressing dispersion, n will be given as a single number, regardless of color.

TOTAL INTERNAL REFLECTION

When a beam of light strikes the boundary to a medium that has a lower index of refraction, the beam bends away from the normal. As the angle of incidence increases, the angle of refraction becomes larger. At some point, when the angle of incidence reaches a **critical angle**, θ_c, the angle of refraction becomes 90°, which means the refracted beam is directed along the surface.

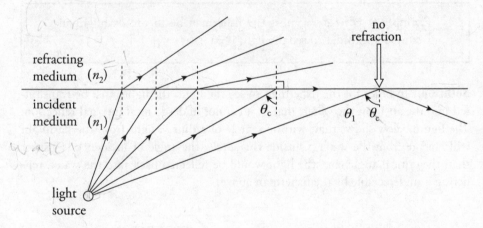

For angles of incidence that are greater than θ_c, there is *no* angle of refraction; the entire beam is reflected back into the original medium. This phenomenon is called **total internal reflection** (sometimes abbreviated **TIR**).

Total internal reflection occurs when:

1) $n_1 > n_2$

and

2) $\theta_1 > \theta_c$, where $\theta_c = \sin^{-1}(n_2/n_1)$

Subscripts
Always think of n_1 or θ_1 as the incidence ray, meaning where the source comes from. Treat n_2 or θ_2 as the resulting ray.

Notice that total internal reflection cannot occur if $n_1 < n_2$. This is because the largest output of $\sin \theta$ is 1, so the largest input of $\sin^{-1}(x)$ is 1. If $n_1 > n_2$, then total internal reflection is a possibility; it will occur if the angle of incidence is large enough, that is, if it's greater than the critical angle, θ_c.

Example 7 What is the critical angle for total internal reflection between air and water? In which of these media must light be incident for total internal reflection to occur?

Solution. First, total internal reflection can only occur when the light is incident in the medium that has the greater refractive index and strikes the boundary to a medium that has a lower index. So, in this case, total internal reflection can occur only when the light source is in the water and the light is incident upon the water/air surface. The critical angle is found as follows:

$$\sin \theta_c = \frac{n_2}{n_1} \quad \Rightarrow \quad \sin \theta_c = \frac{n_{air}}{n_{water}} = \frac{1}{1.33} \quad \Rightarrow \quad \sin \theta_c = 0.75 \quad \Rightarrow \quad \theta_c = 49°$$

Total internal reflection will occur if the light from the water strikes the water/air boundary at an angle of incidence greater than 49°.

Example 8 How close must the fisherman be to the flashlight in Example 5 in order to see the light it emits?

Solution. In order for the fisherman to see the light, the light must be transmitted into the air from the water; that is, it cannot undergo total internal reflection. The figure below shows that, within a circle of radius x, light from the flashlight will emerge from the water. Outside this circle, the angle of incidence is greater than the critical angle, and the light would be reflected back into the water, rendering it undetectable by the fisherman above.

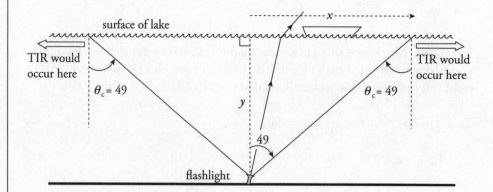

Remember to Check
For optics problems you need to always remember to check if the refracted light ray undergoes total internal reflection. If it undergoes total internal reflection there will be no refracted ray. Hence, if a trick problem asks for an angle of refraction for a siutation that undergoes total internal reflection, there is no angle of refraction.

Because the critical angle for total internal reflection at a water/air interface is 49° (as we found in the preceding example), we can solve for x:

$$\tan 49° = \frac{x}{y} \quad \Rightarrow \quad x = y \tan 49° = (10 \text{ m}) \tan 49° = 11.5 \text{ m}$$

Example 9 The refractive index for the glass prism shown below is 1.55. In order for a beam of light to experience total internal reflection at the right-hand face of the prism, the angle θ_1 must be smaller than what value?

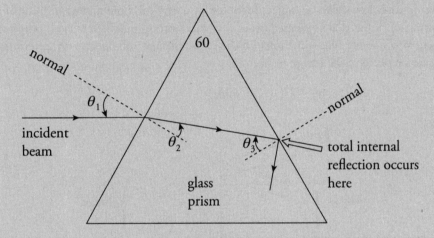

Solution. Total internal reflection will occur at the glass/air boundary if θ_3 is greater than the critical angle, θ_c, which we can calculate this way:

$$\sin \theta_c = \frac{n_{\text{air}}}{n_{\text{glass}}} = \frac{1}{1.55} \quad \Rightarrow \quad \theta_c = 40°$$

Because $\theta_3 = 60° - \theta_2$, total internal reflection will take place if θ_2 is smaller than 20°. Now, by Snell's Law, $\theta_2 = 20°$ if

$$n_{\text{air}} \sin \theta_1 = n_{\text{glass}} \sin \theta_2$$
$$\sin \theta_1 = 1.55 \sin 20°$$
$$\theta_1 = 32°$$

Therefore, total internal reflection will occur at the right-hand face of the prism if θ_1 is smaller than 32°.

MIRRORS

A **mirror** is an optical device that forms an image by reflecting light. We've all looked into a mirror and seen images of nearby objects, and the purpose of this section is to analyze these images mathematically. We begin with a plane mirror, which is flat, and is the simplest type of mirror. Then we'll examine curved mirrors; we'll have to use geometrical methods or algebraic equations to analyze the patterns of reflection from these.

Plane Mirrors

The figure below shows an object (denoted by a vertical, bold arrow) in front of a flat mirror. Light that's reflected off of the object strikes the mirror and is reflected back to our eyes. The directions of the rays reflected off the mirror determine where we perceive the image to be.

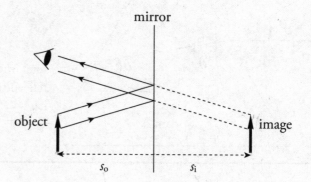

There are four questions we'll answer about the image formed by a mirror:

(1)   Where is the image?

(2)   Is the image real or is it virtual?

(3)   Is the image upright or is it inverted?

(4)   What is the height of the image (compared with that of the object)?

When we look at ourselves in a mirror, it seems like our image is *behind* the mirror, and, if we take a step back, our image also takes a step back. The Law of Reflection can be used to show that the image seems as far behind the mirror as the object is in front of the mirror. This answers question (1).

An image is said to be **real** if light rays actually focus at the image. A real image can be projected onto a screen. For a flat mirror, light rays bounce off the front of the mirror; so, of course, no light focuses behind it. Therefore, the images produced by a flat mirror are not real; they are **virtual**. This answers question (2).

When we look into a flat mirror, our image isn't upside down; flat mirrors produce upright images, and question (3) is answered.

Finally, the image formed by a flat mirror is neither magnified nor diminished (minified) relative to the size of the object. This answers question (4).

Spherical Mirrors

A **spherical mirror** is a mirror that's curved in such a way that its surface forms part of a sphere.

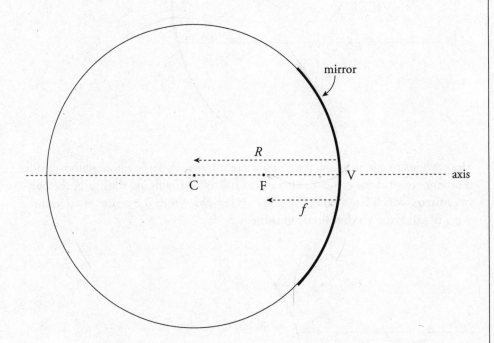

The center of this imaginary sphere is the mirror's **center of curvature**, and the radius of the sphere is called the mirror's **radius of curvature**, R. Halfway between the mirror and the center of curvature, C, is the **focus** (or **focal point**), F. The intersection of the mirror's optic **axis** (its axis of symmetry) with the mirror itself is called the **vertex**, V, and the distance from V to F is called the **focal length**, f, equal to one-half of the radius of curvature:

$$f = \frac{R}{2}$$

If the mirror had a parabolic cross-section, then any ray parallel to the axis would be reflected by the mirror through the focal point. Spherical mirrors do this for incident light rays near the axis (**paraxial rays**) because, in the region of the mirror that's close to the axis, the shapes of a parabolic mirror and a spherical mirror are nearly identical.

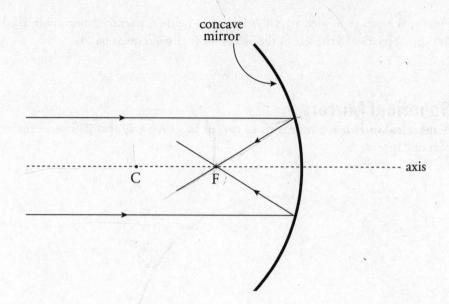

The previous two figures illustrate a **concave mirror**, a mirror whose reflective side is *caved in* toward the center of curvature. The following figure illustrates the **convex mirror**, which has a reflective side curving away from the center of curvature. We will call F the **virtual focus** in this case.

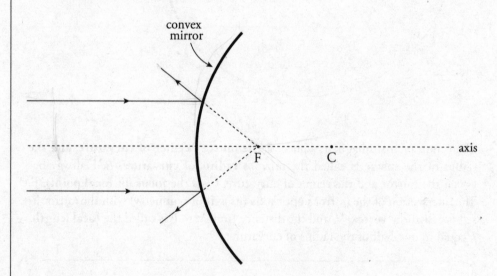

RAY TRACING FOR MIRRORS

One method of answering the four questions listed above concerning the image formed by a mirror involves a geometric approach called **ray tracing**. Representative rays of light are sketched in a diagram that depicts the object and the mirror; the point at which the reflected rays intersect (or appear to intersect) is the location of the image. Some rules governing rays are:

Concave mirrors:

- An incident ray parallel to the axis is reflected through the focus.
- An incident ray that passes through the focus is reflected parallel to the axis.
- An incident ray that strikes the vertex is reflected at an equal angle to the axis.

Convex mirrors:

- An incident ray parallel to the axis is reflected away from the virtual focus.
- An incident ray directed toward the virtual focus is reflected parallel to the axis.
- An incident ray that strikes the vertex is reflected at an equal angle to the axis.

Example 10 The figure below shows a concave mirror and an object (the bold arrow). Use a ray diagram to locate the image of the object.

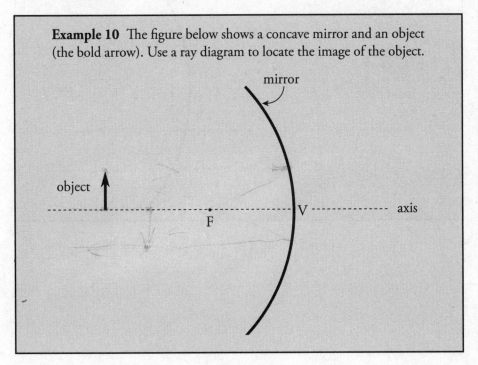

Solution. It only takes two distinct rays to locate the image:

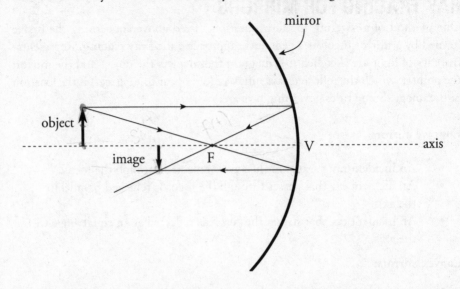

Example 11 The figure below shows a convex mirror and an object (the arrow). Use a ray diagram to locate the image of the object.

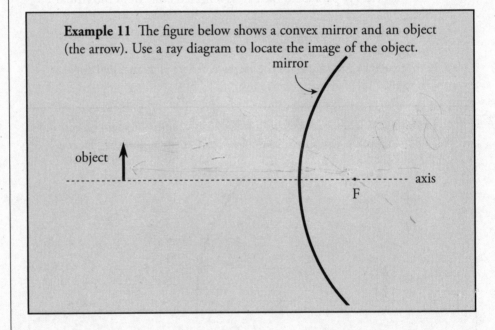

Solution.

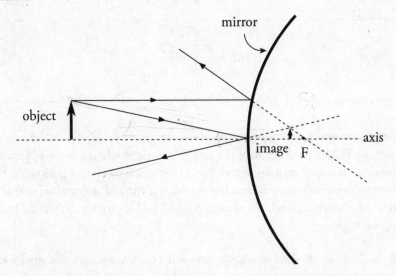

The ray diagrams of the preceding examples can be used to determine the location, orientation, and size of the image. The nature of the image—that is, whether it's real or virtual—can be determined by seeing on which side of the mirror the image is formed. If the image is formed on the same side of the mirror as the object, then the image is real, but if the image is formed on the opposite side of the mirror, it's virtual. Another way of looking at this is: if you had to trace lines back to form an image, that image is virtual. Therefore, the image in Example 10 is real, and the image in Example 11 is virtual.

Using Equations to Answer Questions About the Image

While ray diagrams can answer our questions about images completely, the fastest and easiest way to get information about an image is to use two equations and some simple conventions. The first equation, called the **mirror equation**, is

$$\frac{1}{s_o} + \frac{1}{s_i} = \frac{1}{f}$$

Equation Sheet

where s_o is the object's distance from the mirror, s_i is the image's distance from the mirror, and f is the focal length of the mirror. The value of s_o is *always* positive for a real object, but s_i can be positive or negative. The sign of s_i tells us whether the image is real or virtual: If s_i is positive, the image is real; and if s_i is negative, the image is virtual.

The second equation is called the **magnification equation**:

$$M = \frac{h_i}{h_o} = \frac{-s_i}{s_o}$$

This gives the magnification; the height of the image, h_i, is $|M|$ times the height of the object, h_o. If M is positive, then the image is upright relative to the object; if M is negative, it's inverted relative to the object. Because s_o is always positive, we can come to two important conclusions. If s_i is positive, then M is negative, so real images are always inverted, and, if s_i is negative, then M is positive, so virtual images are always upright.

Finally, to distinguish *mathematically* between concave and convex mirrors, we always write the focal length f as a positive value for concave mirrors and a negative value for convex mirrors. With these two equations and their accompanying conventions, all four questions about an image can be answered.

MIRRORS

converging—concave $\quad\Leftrightarrow\quad f$ positive

diverging—convex $\quad\Leftrightarrow\quad f$ negative

$$\frac{1}{s_o} + \frac{1}{s_i} = \frac{1}{f}$$
$\begin{cases} s_o \text{ always positive (real object)} \\ s_i \text{ positive} \quad\Rightarrow\quad \text{image is real (located on} \\ \qquad\qquad\qquad\qquad \text{the } \textit{same} \text{ side of mirror as object)} \\ s_i \text{ negative} \quad\Rightarrow\quad \text{image is virtual (located on} \\ \qquad\qquad\qquad\qquad \textit{opposite} \text{ side of mirror from object)} \end{cases}$

$$M = \frac{h_i}{h_o} = \frac{-s_i}{s_o}$$
$\begin{cases} \text{given } h_o \text{ is positive} \\ h_i \text{ and } M \text{ positive} \quad\Rightarrow\quad \text{image is upright} \\ \\ h_i \text{ and } M \text{ negative} \quad\Rightarrow\quad \text{image is inverted} \end{cases}$

Which is the Positive Side?

We will always treat the object's position s_o as being the positive side. For mirrors, if the image is produced on the same side as the object it is considered a positive distance. If an image is produced opposite the side of the object it is considered a negative distance. Concave mirrors will always have a positive focal length (meaning the focal point is on the same side of the object). Convex mirrors will always have a negative focal length (meaning the focal point is on the opposite side of the object).

Real or Virtual?

Remember that all real images are inverted and all virtual images are upright.

Example 12 An object of height 4 cm is placed 30 cm in front of a concave mirror whose focal length is 10 cm.

        (a) Where's the image?

        (b) Is it real or virtual?

        (c) Is it upright or inverted?

        (d) What's the height of the image?

Solution.

(a)   With s_o = 30 cm and f = 10 cm, the mirror equation gives

$$\frac{1}{s_o} + \frac{1}{s_i} = \frac{1}{f} \Rightarrow \frac{1}{30 \text{ cm}} + \frac{1}{s_i} = \frac{1}{10 \text{ cm}} \Rightarrow \frac{1}{s_i} = \frac{1}{15 \text{ cm}} \Rightarrow s_i = 15 \text{ cm}$$

The image is located 15 cm in front of the mirror.

(b)   Because s_i is positive, the image is real.

(c)   Real images are inverted.

(d) $\frac{h_i}{h_o} = \frac{-s_i}{s_o} \Rightarrow h_i = \frac{-s_i h_o}{s_o} \Rightarrow h_i = \frac{(-15 \text{ cm})(4 \text{ cm})}{30 \text{ cm}} = -2 \text{ cm}$

The −2 cm also confirms the image is inverted.

Example 13 An object of height 4 cm is placed 20 cm in front of a convex mirror whose focal length is −30 cm.

        (a) Where's the image?

        (b) Is it real or virtual?

        (c) Is it upright or inverted?

        (d) What's the height of the image?

Solution.

(a)   With s_o = 20 cm and f = −30 cm, the mirror equation gives us:

$$\frac{1}{s_o} + \frac{1}{s_i} = \frac{1}{f} \Rightarrow \frac{1}{20 \text{ cm}} + \frac{1}{s_i} = \frac{1}{-30 \text{ cm}} \Rightarrow \frac{1}{s_i} = -\frac{1}{12 \text{ cm}} \Rightarrow s_i = -12 \text{ cm}$$

So, the image is located 12 cm behind the mirror.

(b)   Because s_i is negative, the image is virtual.

(c)   Virtual images are upright.

What Produces What?

Only concave mirrors can produce both real images (if $s_o > f$) and virtual images (if $s_o < f$). Convex mirrors and plane mirrors can only produce virtual images.

(d) $\dfrac{h_i}{h_o} = \dfrac{-s_i}{s_o} \Rightarrow h_i = \dfrac{-s_i h_o}{s_o} \Rightarrow h_i = \dfrac{-(-12 \text{ cm})(4 \text{ cm})}{20 \text{ cm}} = 2.4 \text{ cm}$

[The fact that the magnification is positive tells us that the image is upright, as we said in part (c).]

Example 14 Show how the statements made earlier about plane mirrors can be derived from the mirror and magnification equations.

Solution. A plane mirror can be considered a spherical mirror with an infinite radius of curvature (and an infinite focal length). If $f = \infty$, then $1/f = 0$, so the mirror equation becomes

$$\frac{1}{s_o} + \frac{1}{s_i} = 0 \quad \Rightarrow \quad s_i = -s_o$$

So, the image is as far behind the mirror as the object is in front. Also, since s_o is always positive, s_i is negative, so the image is virtual. The magnification is

$$M = -\frac{s_i}{s_o} = -\frac{-s_o}{s_o} = 1$$

and the image is upright and has the same height as the object. The mirror and magnification equations confirm our description of images formed by plane mirrors.

Example 15 Show why convex mirrors can only form virtual images.

Solution. Because f is negative and s_o is positive, the mirror equation

$$\frac{1}{s_o} + \frac{1}{s_i} = \frac{1}{f}$$

immediately tells us that s_i cannot be positive (if it were, the left-hand side would be the sum of two positive numbers, while the right-hand side would be negative). Since s_i must be negative, the image must be virtual.

Example 16 An object placed 60 cm in front of a spherical mirror forms a real image at a distance of 30 cm from the mirror.

      (a) Is the mirror concave or convex?

      (b) What's the mirror's focal length?

      (c) Is the image taller or shorter than the object?

Solution.

(a)   The fact that the image is real tells us that the mirror cannot be convex, since convex mirrors form only virtual images. The mirror is concave.

(b)   With $s_o = 60$ cm and $s_i = 30$ cm (s_i is positive since the image is real), the mirror equation tells us that

$$\frac{1}{s_o} + \frac{1}{s_i} = \frac{1}{f} \Rightarrow \frac{1}{60 \text{ cm}} + \frac{1}{30 \text{ cm}} = \frac{1}{f} \Rightarrow \frac{1}{f} = \frac{1}{20 \text{ cm}} \Rightarrow f = 20 \text{ cm}$$

Note that f is positive, which confirms that the mirror is concave.

(c)   The magnification is

$$M = -\frac{s_i}{s_o} = -\frac{30 \text{ cm}}{60 \text{ cm}} = -\frac{1}{2}$$

Since the absolute value of M is less than 1, the mirror makes the object look smaller (minifies the height of the object). The image is only half as tall as the object (and is upside down, because M is negative).

Example 17 A concave mirror with a focal length of 25 cm is used to create a real image that has twice the height of the object. How far is the image from the mirror?

Solution. Since h_i (the height of the image) is twice h_o (the height of the object) the value of the magnification is either +2 or –2. To figure out which, we just notice that the image is real; real images are inverted, so the magnification, M, must be negative. Therefore, $M = -2$, so

$$-\frac{s_i}{s_o} = -2 \Rightarrow s_o = \frac{1}{2} s_i$$

Substituting this into the mirror equation gives us

$$\frac{1}{s_o} + \frac{1}{s_i} = \frac{1}{f} \Rightarrow \frac{1}{\frac{1}{2}s_i} + \frac{1}{s_i} = \frac{1}{f} \Rightarrow \frac{3}{s_i} = \frac{1}{f} \Rightarrow s_i = 3f = 3(25 \text{ cm}) = 75 \text{ cm}$$

THIN LENSES

A lens is an optical device that forms an image by *refracting* light. We'll now talk about the equations and conventions that are used to analyze images formed by the two major categories of lenses: converging and diverging.

A **converging lens**—like the bi-convex one shown below—converges parallel paraxial rays of light to a focal point on the far side. (This lens is *bi-convex*; both of its faces are convex. Converging lenses all have at least one convex face.) Because parallel light rays actually focus at F, this point is called a **real focus**. Its distance from the lens is the focal length, *f*.

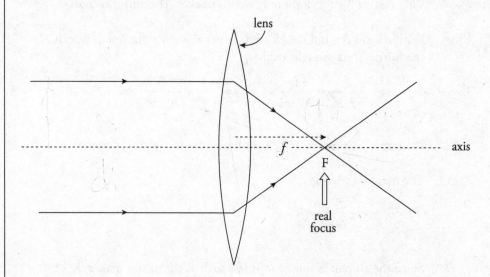

So Many Equations!
The purpose of a mirror is to reflect an image. The purpose of a lens is to refract an image (or see through it). Fortunately, the equations for mirrors and lenses will be the same with a minor difference on the focal length.

A **diverging lens**—like the *bi-concave* one shown below—causes parallel paraxial rays of light to diverge away from a **virtual focus**, F, on the same side as the incident rays. (Diverging lenses all have at least one concave face.)

Don't Mix It Up
Converging Lens = Bi-Convex Lens
Diverging Lens = Bi-Concave Lens

RAY TRACING FOR LENSES

Just as is the case with mirrors, representative rays of light can be sketched in a diagram along with the object and the lens; the point at which the reflected rays intersect (or appear to intersect) is the location of the image. The rules that govern these rays are as follows:

Converging lenses:

- An incident ray parallel to the axis is refracted through the real focus.

- Incident rays pass undeflected through the **optical center**, O (the central point within the lens where the axis intersects the lens).

Diverging lenses:

- An incident ray parallel to the axis is refracted away from the virtual focus.

- Incident rays pass undeflected through the optical center, O.

Example 18 The figure below shows a converging lens and an object (denoted by the bold arrow). Use a ray diagram to locate the image of the object.

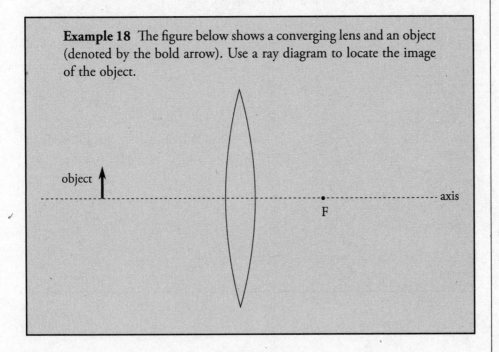

Solution.

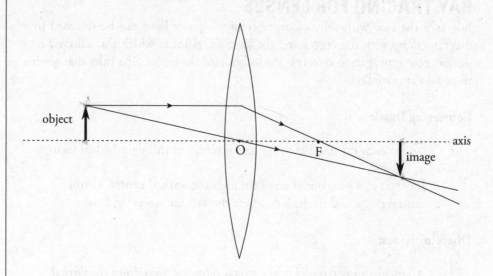

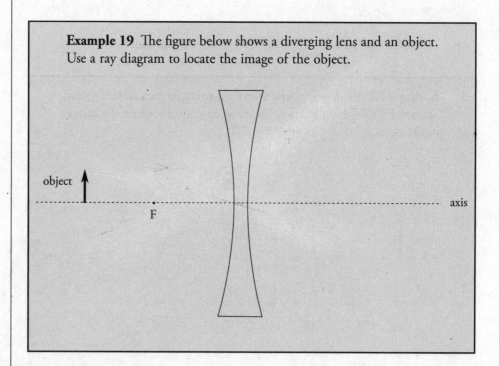

Example 19 The figure below shows a diverging lens and an object. Use a ray diagram to locate the image of the object.

Solution.

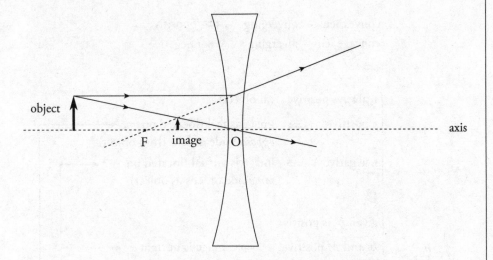

The nature of the image—that is, whether it's real or virtual—is determined by the side of the lens upon which the image is formed. If the image is formed on the side of the lens that's opposite the object, then the image is real, and if the image is formed on the same side of the lens as the object, then it's virtual. Another way of looking at this is: If you had to trace lines back to form an image, that image is virtual. Therefore, the image in Example 18 is real, while the image in Example 19 is virtual.

USING EQUATIONS TO ANSWER QUESTIONS ABOUT THE IMAGE

Lenses and mirrors use the same equations, notation, and sign conventions, with the following note. Converging optical devices ($+f$) are con<u>cave</u> mirrors and con<u>vex</u> lenses. Diverging optical devices ($-f$) are con<u>vex</u> mirrors and con<u>cave</u> lenses.

LENSES

convex lens—converging ⟺ f positive

concave lens—diverging ⟺ f negative

$$\frac{1}{s_o} + \frac{1}{s_i} = \frac{1}{f} \begin{cases} s_o \text{ always positive (real object)} \\ s_i \text{ positive} \Rightarrow \text{image is real (located on } opposite \text{ side of lens from object)} \\ s_i \text{ negative} \Rightarrow \text{image is virtual (located on } same \text{ side of lens as object)} \end{cases}$$

$$M = \frac{h_i}{h_o} = \frac{-s_i}{s_o} \begin{cases} \text{given } h_o \text{ is positive} \\ h_i \text{ and } M \text{ positive} \Rightarrow \text{image is upright} \\ h_i \text{ and } M \text{ negative} \Rightarrow \text{image is inverted} \end{cases}$$

Example 20 An object of height 11 cm is placed 44 cm in front of a converging lens with a focal length of 24 cm.

    (a) Where's the image?

    (b) Is it real or virtual?

    (c) Is it upright or inverted?

    (d) What's the height of the image?

Solution.

(a)   With $s_o = 44$ cm and $f = 24$ cm, the lens equation gives us:

$$\frac{1}{s_o} + \frac{1}{s_i} = \frac{1}{f} \Rightarrow \frac{1}{44 \text{ cm}} + \frac{1}{s_i} = \frac{1}{24 \text{ cm}} \Rightarrow \frac{1}{s_i} = 0.0189 \text{ cm}^{-1} \Rightarrow s_i = 53 \text{ cm}$$

So, the image is located 53 cm from the lens, on the opposite side from the object.

(b)   Because s_i is positive, the image is real.

(c)   Real images are inverted.

(d) $\dfrac{h_i}{h_o} = \dfrac{-s_i}{s_o} \Rightarrow h_i = \dfrac{-s_i h_o}{s_o} \Rightarrow h_i = \dfrac{(-53 \text{ cm})(11 \text{ cm})}{44 \text{ cm}} = -13 \text{ cm}$

The negative h_i reaffirms that we have an inverted image.

Example 21 An object of height 11 cm is placed 48 cm in front of a diverging lens with a focal length of –24.5 cm.

   (a) Where's the image?

   (b) Is it real or virtual?

   (c) Is it upright or inverted?

   (d) What's the height of the image?

Solution.

(a)   With s_o = 48 cm and f = –24.5 cm, the lens equation gives us:

$$\frac{1}{s_o} + \frac{1}{s_i} = \frac{1}{f} \quad \Rightarrow \quad \frac{1}{48 \text{ cm}} + \frac{1}{s_i} = \frac{1}{-24.5 \text{ cm}} \quad \Rightarrow \quad \frac{1}{s_i} = -0.0616 \text{ cm}^{-1} \quad \Rightarrow \quad s_i = -16 \text{ cm}$$

The image is 16 cm from the lens, on the same side as the object.

(b)   Because s_i is negative, the image is virtual.

(c)   Virtual images are upright.

(d) $\frac{h_i}{h_o} = \frac{-s_i}{s_o} \Rightarrow h_i = \frac{-s_i h_o}{s_o} \Rightarrow h_i = \frac{-(-16 \text{ cm})(11 \text{ cm})}{48 \text{ cm}} = 3.7 \text{ cm}$

These results are illustrated in Example 19.

What Produces What?
Only converging lenses can produce both real images (if $s_0 > f$) and virtual images (if $s_0 < f$). Diverging lenses can only produce virtual images.

Example 22 The **power** of a lens, P, is defined as $1/f$, where f is the lens's focal length (in meters). Lens power is expressed in **diopters** (abbreviated D), where 1 D = 1 m^{-1}. When an object is placed 10 cm from a converging lens, the real image formed is twice the height of the object. What's the power of this lens?

Solution. Since h_i, the height of the image, is twice h_o, the height of the object, the value of the magnification is either +2 or –2. To decide, we simply notice that the image is real; real images are inverted, so the magnification, m, must be negative. Since $m = -2$, we find that

$$-\frac{s_i}{s_o} = -2 \quad \Rightarrow \quad s_o = \frac{1}{2}s_i$$

Because $s_o = 10$ cm, it must be true that $s_i = 20$ cm, so the lens equation gives us:

$$\frac{1}{s_o} + \frac{1}{s_i} = \frac{1}{f} \quad \Rightarrow \quad \frac{1}{10 \text{ cm}} + \frac{1}{20 \text{ cm}} = \frac{1}{f} \quad \Rightarrow \quad f = \frac{20}{3} \text{ cm}$$

Expressing the focal length in meters and then taking the reciprocal gives the lens power in diopters:

$$f = \frac{20}{3} \text{ cm} \times \frac{1 \text{ m}}{100 \text{ cm}} = \frac{1}{15} \text{ m} \quad \Rightarrow \quad P = \frac{1}{f} = \frac{1}{\frac{1}{15} \text{ m}} = 15 \text{ D}$$

Chapter 10 Review Questions

Solutions can be found in Chapter 12.

Section I: Multiple Choice

1. What is the wavelength of an X-ray whose frequency is 1.0×10^{18} Hz ?

   (A) 3.3×10^{-11} m
   (B) 3.0×10^{-10} m
   (C) 3.3×10^{-9} m
   (D) 3.0×10^{-8} m

2. In Young's double-slit interference experiment, what is the difference in path length of the light waves from the two slits at the center of the first bright fringe above the central maximum?

   (A $\frac{1}{4}\lambda$

   (B) $\frac{1}{2}\lambda$

   (C) λ

   (D) $\frac{3}{2}\lambda$

3. A beam of light in air is incident upon the smooth surface of a piece of flint glass, as shown:

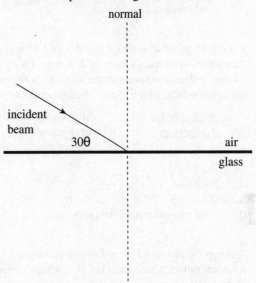

   If the reflected beam and refracted beam are perpendicular to each other, what is the index of refraction of the glass?

   (A) $\frac{1}{2}$

   (B) $\frac{1}{2}\sqrt{3}$

   (C) $\sqrt{3}$

   (D) $2\sqrt{3}$

4. When green light (wavelength = 500 nm in air) travels through diamond (refractive index = 2.5), what is its wavelength?

   (A) 200 nm
   (B) 300 nm
   (C) 500 nm
   (D) 1250 nm

5. A beam of light traveling in Medium 1 strikes the interface to another transparent medium, Medium 2. If the speed of light is less in Medium 2 than in Medium 1, the beam will

(A) refract toward the normal
(B) refract away from the normal
(C) undergo total internal reflection
(D) have an angle of reflection smaller than the angle of incidence

6. If a clear liquid has a refractive index of 1.45 and a transparent solid has an index of 2.90 then, for total internal reflection to occur at the interface between these two media, which of the following must be true?

incident beam originates in	at an angle of incidence greater than
(A) The solid	30°
(B) The liquid	30°
(C) The liquid	60°
(D) Total internal reflection cannot occur.	

7. An object is placed 60 cm in front of a concave spherical mirror whose focal length is 40 cm. Which of the following best describes the image?

	Nature of image	Distance from mirror
(A)	Virtual	24 cm
(B)	Real	24 cm
(C)	Virtual	120 cm
(D)	Real	120 cm

8. An object is placed 60 cm from a spherical convex mirror. If the mirror forms a virtual image 20 cm from the mirror, what's the magnitude of the mirror's radius of curvature?

(A) 7.5 cm
(B) 15 cm
(C) 30 cm
(D) 60 cm

9. The image created by a converging lens is projected onto a screen that's 60 cm from the lens. If the height of the image is 1/4 the height of the object, what's the focal length of the lens?

(A) 36 cm
(B) 45 cm
(C) 48 cm
(D) 72 cm

10. Which of the following is true concerning a bi-concave lens? (A bi-concave lens has both surfaces concave.)

(A) Its focal length is positive.
(B) It cannot form real images.
(C) It cannot form virtual images.
(D) It can magnify objects.

Section II: Free Response

1. Two trials of a double-slit interference experiment are set up as follows. The slit separation is $d = 0.50$ mm, and the distance to the screen, L, is 4.0 m.

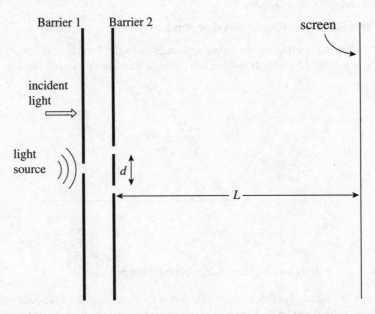

(a) What is the purpose of the first (single-slit) barrier? Why not use two light sources, one at each slit at the second barrier? Explain briefly.

In the first trial, white light is used.

(b) What is the vertical separation on the screen (in mm) between the first-order maxima for red light ($\lambda = 750$ nm) and violet light ($\lambda = 400$ nm)?

(c) Locate the nearest point to the central maximum where an intensity maximum for violet light ($\lambda = 400$ nm) coincides with an intensity maximum for orange-yellow light ($\lambda = 600$ nm).

In the second trial, the entire region between the double-slit barrier and the screen is filled with a large slab of glass of refractive index $n = 1.5$, and monochromatic green light ($\lambda = 500$ nm in air) is used.

(d) What is the separation between adjacent bright fringes on the screen?

2. In order to determine the criteria for constructive and destructive interference, the following rules are used:

   i) When light strikes the boundary to a medium with a higher refractive index than the incident medium, it undergoes a 180° phase change upon reflection (this is equivalent to a shift by one-half wavelength).

   ii) When light strikes the boundary to a medium with a lower refractive index than the incident medium, it undergoes no phase change upon reflection.

   These rules can be applied to the two situations described below.

   A thin soap film of thickness T, consisting of a mixture of water and soap (refractive index = 1.38), has air on both sides. Incident sunlight is reflected off the front face and the back face, causing interference.

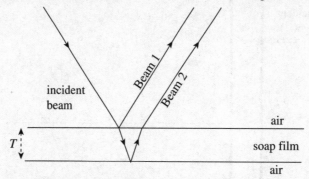

   (a) Which beam, 1 or 2, suffers a 180° phase change upon reflection?

   (b) Since the beams are out of phase, destructive interference will occur if the difference in their path lengths, $\Delta\ell \approx 2T$ for near-normal incidence, is equal to a whole number of wavelengths (wavelength as measured in the soap film). What is the criterion for constructive interference? Write your answer as an algebraic equation.

3. An object of height 5 cm is placed 40 cm in front of a spherical concave mirror. An image is formed 72 cm behind the mirror.

    (a)   Is the image real or virtual?

    (b)   Is the image upright or inverted?

    (c)   What's the height of the image?

    (d)   What is the mirror's radius of curvature?

    (e)   In the figure below, sketch the mirror, labeling its vertex and focal point, and then construct a ray diagram (with a minimum of two rays) showing the formation of the image.

Summary

○ All electromagnetic waves travel at the speed of light ($c = 3 \times 10^8$ m/s). They obey the wave equation $v = f\lambda$. When monochromatic coherent light goes through double slits (or a diffraction grating) there will be constructive interference when the path difference is given by $PD = m\lambda$ or $d \sin \theta = m\lambda$ where $m = 0, 1, 2 \ldots$. There will be destructive interference for $PD = (m+\frac{1}{2})\lambda$ or $d \sin \theta = (m+\frac{1}{2})\lambda$. For small angles we can say $x_m = \dfrac{m\lambda L}{d}$, where x_m is the distance the mth maximum is located from the central maximum and shows where constructive interference occurs.

○ Geometric optics always measures angles from the normal line. Topics include the fact that the angle of incidence is equal to the angle of reflection and that these angles lie in the same plane.

o   Snell's Law states when light enters a new medium (material) it changes speeds and may change directions. This is stated in the formula

$$n_1 \sin\theta_1 = n_2 \sin\theta_2$$

where n is the index of refraction, which is a ratio of the speed of light in a vacuum to the speed of light in the substance ($n = \dfrac{c}{v}$). The index of refraction is always greater than 1 and has no units. When going from a high index of refraction to a lower index of refraction, light may have total internal reflection if the angle is larger than the critical angle ($\sin\theta_c = \dfrac{n_2}{n_1}$).

o   For both thin lenses and curved mirrors you can use the thin lens equation:

$$\frac{1}{s_o} + \frac{1}{s_i} = \frac{1}{f}$$

Where f is positive for a convex lens or a concave mirror and f is negative for a concave lens or a convex mirror s_0 is always positive.

o   For a lens, if s_i is negative the lens image is virtual and located on the same side of the lens as the object. If s_i is positive, the lens image is real and located on the opposite side of the lens from the object.

o   For a mirror, if s_i is negative, the image is virtual and located on the opposite side of the mirror from the object. If s_i is positive, the image is real and located on the same side of the mirror as the object.

o The magnification of a lens is given by
$$M = \frac{h_i}{h_o} = \frac{-s_i}{s_o}.$$

o A **diverging optical device** could be either a convex mirror or a concave lens. The thing these two have in common is that no matter where the object is located they form images that are virtual and upright. The sign of h_i will always be positive and the sign of s_i will always be negative. The sign of the magnification will be positive and its value will be less than one. The image will always be located between the focal length and the optical device.

o A **converging optical device** could be either a concave mirror or a convex lens. The thing these two have in common is that when s_o is *outside* the focal length they form images that are real and inverted. The sign of h_i will be negative and the sign of s_i will be positive. The sign of the magnification will be negative and its absolute value can be any number greater than zero. The other thing they have in common is that when s_o is *inside* the focal length they form images that are virtual and upright. The sign of h_i will be positive and the sign of s_i will be negative. The sign of the magnification will be positive and its value will be greater than one.

Chapter 11
Modern Physics

INTRODUCTION

The subject matter of the previous chapters was developed in the seventeenth, eighteenth, and nineteenth centuries, but as we delve into the physics of the very small, we enter the twentieth century.

PHOTONS AND THE PHOTOELECTRIC EFFECT

Max Planck first proposed the idea of light being emitted as individual packets of constant energy, called **quanta**. This is where the name quantum mechanics comes from. The particle nature of light was pioneered by Einstein in 1905, when he showed light transferred energy like a particle, and Arthur Compton in 1923, when he showed light has momentum and can undergo elastic collisions. A quantum of electromagnetic energy is known as a **photon**. Light behaves like a stream of photons, and this is illustrated by the **photoelectric effect**.

When a piece of metal is illuminated by electromagnetic radiation (specifically visible light, ultraviolet light, or X-rays), the energy absorbed by electrons near the surface of the metal can liberate them from their bound state, and these electrons can fly off. The released electrons are known as **photoelectrons**. In this case, the classical, wave-only theory of light would predict three results:

(1)   There would be a significant time delay between the moment of illumination and the ejection of photoelectrons. This is because the metal would have to heat up to the point that the thermal energy was enough to overcome the binding energy.

(2)   Increasing the intensity of the light would cause the electrons to leave the metal surface with greater kinetic energy. This is because the energy carried by a wave is related to its intensity.

(3)   Photoelectrons would be emitted regardless of the frequency of the incident energy, as long as the intensity was high enough.

Surprisingly, none of these predictions was observed. Photoelectrons were ejected within just a few billionths of a second after illumination, disproving prediction (1). Secondly, increasing the intensity of the light did not cause photoelectrons to leave the metal surface with greater kinetic energy. Although more electrons were ejected as the intensity was increased, there was a maximum photoelectron kinetic energy; prediction (2) was false. And, for each metal, there was a certain **threshold frequency**, f_0: If light of frequency lower than f_0 were used to illuminate the metal surface, *no* photoelectrons were ejected, regardless of how intense the incident radiation was; prediction (3) was also false. Clearly, something was wrong with the wave-only theory of light.

Einstein explained these observations by borrowing from Planck's idea that light came in individual quanta of energy. He called the quanta photons. The energy of a photon is proportional to the frequency of the wave,

Equation Sheet

$$E = hf$$

where h is **Planck's constant** (about 6.63×10^{-34} J·s). A certain amount of energy had to be imparted to an electron on the metal surface in order to liberate it; this was known as the metal's **work function**, or ϕ. If an electron absorbed a photon whose energy E was greater than ϕ, it would leave the metal with a maximum kinetic energy equal to $E - \phi$. This process could occur *very* quickly, which accounts for the rapidity with which photoelectrons are produced after illumination. This explains why prediction (1) was incorrect.

In this view, increasing the intensity (and therefore the energy) just means increasing the number of photons and results in the ejection of more photoelectrons—but since the energy of each incident photon is fixed by the equation $E = hf$, the value of K_{max} will still be $E - \phi$. This can be expressed as

Equation Sheet

$$K_{max} = hf - \phi$$

This accounts for the observation that disproved prediction (2).

Finally, if the incoming photon's energy was less than phi (or, $E = hf < \phi$), the photon energy would not be enough to liberate electrons. Blasting the metal surface with more photons (that is, increasing the intensity of the incident beam) would also do nothing; none of the photons would have enough energy to eject electrons. This accounts for the observation of a threshold frequency, which we now know is ϕ/h. This can be expressed as

$$f_o = \phi/h$$

and is why prediction (3) was incorrect. Before we get to some examples, it's worthwhile to introduce a new unit for energy. The SI unit for energy is the joule, but it's too large to be convenient in the domains we're studying now. We'll use a much smaller unit, the **electronvolt** (abbreviated **eV**). The eV is equal to the energy gained (or lost) by an electron accelerated through a potential difference of one volt. Using the equation $\Delta U_E = qV$, we find that

$$1 \text{ eV} = (1 \text{ e})(1 \text{ V}) = (1.6 \times 10^{-19} \text{ C})(1 \text{ V}) = 1.6 \times 10^{-19} \text{ J}$$

In terms of electronvolts, the value of Planck's constant is 4.14×10^{-15} eV·s.

Example 1 The work function, ϕ, for aluminum is 4.08 eV.

(a) What is the threshold frequency required to produce photoelectrons from aluminum?

(b) Classify the electromagnetic radiation that can produce photoelectrons.

(c) If light of frequency $f = 4.00 \times 10^{15}$ Hz is used to illuminate a piece of aluminum,

    (i) what is K_{max}, the maximum kinetic energy of ejected photoelectrons?

    (ii) what's the maximum speed of the photoelectrons? (Electron mass = 9.11×10^{-31} kg.)

(d) If the light described in part (b) were increased by a factor of 2 in intensity, what would happen to the value of K_{max}?

Solution.

(a) We know from the statement of the question that, in order for a photon to be successful in liberating an electron from the surface of the aluminum, its energy cannot be less than 4.08 eV. Therefore, the minimum frequency of the incident light—the threshold frequency—must be

$$f_0 = \frac{\phi}{h} = \frac{4.08 \text{ eV}}{4.14 \times 10^{-15} \text{ eV} \cdot \text{s}} = 9.86 \times 10^{14} \text{ Hz}$$

(b) Based on the electromagnetic spectrum given in the previous chapter, the electromagnetic radiation used to produce photoelectrons from aluminum must be at least in the ultraviolet region of the EM spectrum.

(c) (i) The maximum kinetic energy of photoelectrons is found from the equation

$$\begin{aligned} K_{max} &= hf - \phi \\ &= (4.14 \times 10^{-15} \text{ eV} \cdot \text{s})(4.00 \times 10^{15} \text{ Hz}) - (4.08 \text{ eV}) \\ &= 12.5 \text{ eV} \end{aligned}$$

(ii) Using the above result and $K = \frac{1}{2}mv^2$, we can find v_{max}:

$$v_{max} = \sqrt{\frac{2}{m_e}K_{max}}$$

$$= \sqrt{\frac{2}{9.11 \times 10^{-31} \text{ kg}}\; 12.5 \text{ eV} \cdot \frac{1.6 \times 10^{-19} \text{ J}}{1 \text{ eV}}}$$

$$= 2.1 \times 10^6 \text{ m/s}$$

(d)   Nothing will happen. Increasing the intensity of the illuminating radiation will cause more photons to impinge on the metal surface, thereby ejecting more photoelectrons, but their maximum kinetic energy will remain the same. The only way to increase K_{max} would be to increase the frequency of the incident energy.

THE BOHR MODEL OF THE ATOM

In the years immediately following Rutherford's announcement of his nuclear model of the atom, a young physicist, Niels Bohr, would add an important piece to the atomic puzzle. Rutherford told us where the positive charge of the atom was located; Bohr would tell us about the electrons.

For fifty years it had been known that atoms in a gas discharge tube emit and absorb light only at specific wavelengths. The light from a glowing gas, passed through a prism to disperse the beam into its component wavelengths, produces patterns of sharp lines called **atomic spectra**. The visible wavelengths that appear in the emission spectrum of hydrogen had been summarized by the *Balmer formula*

$$\frac{1}{\lambda_n} = R\;\left(\frac{1}{2^2} - \frac{1}{n^2}\right)$$

where R is the *Rydberg constant* (about 1.1×10^7 m^{-1}). The formula worked—that is, it fit the observational data—but it had no theoretical basis. So, the question was, *why* do atoms emit (or absorb) radiation only at certain discrete wavelengths?

Bohr's model of the atom explains the spectroscopists' observations. Using the simplest atom, hydrogen (with only one electron), Bohr postulated that the electron would orbit the nucleus only at certain discrete radii. When the electron is in one of these special orbits, it does not radiate away energy (as the classical theory would predict). However, if the electron absorbs a certain amount of energy, it is **excited** to a higher orbit, one with a greater radius. After a short time in this excited state, it returns to a lower orbit, emitting a photon in the process. Since each allowed orbit—or **energy level**—has a specific radius (and corresponding energy), the photons emitted in each jump have only specific wavelengths.

Chemistry
Modern Physics begins to explore atoms, which bridges into chemistry. Fortunately, you will not need to know much about chemistry for the AP Physics test—just a basic understanding of some topics.

When an excited electron drops from energy level $n = j$ to a lower one, $n = i$, the transition causes a photon of energy to be emitted, and the energy of the photon is the difference between the two energy levels:

$$E_{\text{emitted photon}} = |\ E| = E_j - E_i$$

The wavelength of this photon is

Speed of Light
Photons travel at the speed of light.

$$\lambda = \frac{c}{f} = \frac{c}{E_{\text{photon}}/h} = \frac{hc}{E_j - E_i}$$

Example 2
The first five energy levels of an atom are shown in the diagram below:

-3 eV _____ $n = 5$
-4 eV _____ $n = 4$

-7 eV _____ $n = 3$

-15 eV _____ $n = 2$

-62 eV _____ $n = 1$ ground state

(a) If the atom begins in the $n = 3$ level, what photon energies could be emitted as it returns to its ground state?

(b) What could happen if this atom, while in an undetermined energy state, were bombarded with a photon of energy 10 eV?

Solution.

(a) If the atom is in the $n = 3$ state, it could return to ground state by a transition from $3 \rightarrow 1$, or from $3 \rightarrow 2$ and then $2 \rightarrow 1$. The energy emitted in each of these transitions is simply the difference between the energies of the corresponding levels:

$$E_{3 \rightarrow 1} = E_3 - E_1 = (-7 \text{ eV}) - (-62 \text{ eV}) = 55 \text{ eV}$$

$$E_{3 \rightarrow 2} = E_3 - E_2 = (-7 \text{ eV}) - (-15 \text{ eV}) = 8 \text{ eV}$$

$$E_{2 \rightarrow 1} = E_2 - E_1 = (-15 \text{ eV}) - (-62 \text{ eV}) = 47 \text{ eV}$$

(b)   Since no two energy levels in this atom are separated by 10 eV, the atom could not absorb a 10 eV photon, and as a result, nothing would happen. This atom would be transparent to light of energy 10 eV.

Example 3

(a) Using the energy level diagram for hydrogen below, how much energy must a ground-state electron in a hydrogen atom absorb to be excited to the $n = 4$ energy level?

(b) With the electron in the $n = 4$ level, what wavelengths are possible for the photon emitted when the electron drops to a lower energy level? In what regions of the EM spectrum do these photons lie?

Solution.

(a)   The ground-state energy level ($n = 1$) for hydrogen is -13.6 eV, and $E_4 = -0.85$ eV.

Therefore, in order for an electron to make the transition from E_1 to E_4, it must absorb energy in the amount $E_4 - E_1 = (-0.85 \text{ eV}) - (-13.6 \text{ eV}) = 12.8$ eV.

(b)   An electron in the $n = 4$ energy level can make several different transitions: It can drop to $n = 3$, $n = 2$, or all the way down to the ground state, $n = 1$. The energies of the $n = 2$ and $n = 3$ levels are

$$E_2 = \frac{1}{2^2}(-13.6 \text{ eV}) = -3.4 \text{ eV} \qquad \text{and} \qquad E_3 = \frac{1}{3^2}(-13.6 \text{ eV}) = -1.5 \text{ eV}$$

The following diagram shows the electron dropping from $n = 4$ to $n = 3$:

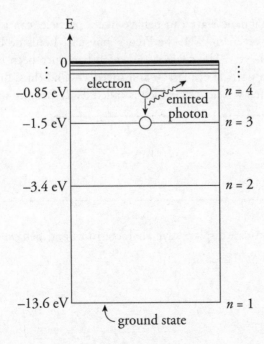

There are three possible values for the energy of the emitted photon, $E_{4 \to 3}$, $E_{4 \to 2}$, or $E_{4 \to 1}$:

$$E_{4 \to 3} = E_4 - E_3 = (-0.85 \text{ eV}) - (-1.5 \text{ eV}) = 0.65 \text{ eV}$$
$$E_{4 \to 2} = E_4 - E_2 = (-0.85 \text{ eV}) - (-3.4 \text{ eV}) = 2.55 \text{ eV}$$
$$E_{4 \to 1} = E_4 - E_1 = (-0.85 \text{ eV}) - (-13.6 \text{ eV}) = 12.8 \text{ eV}$$

From the equation $E = hf = hc/\lambda$, we get $\lambda = hc/E$, so

$$\lambda_{4 \to 3} = \frac{hc}{E_{4 \to 3}} = \frac{(4.14 \times 10^{-15} \text{ eV} \cdot \text{s})(3.00 \times 10^8 \text{ m/s})}{0.65 \text{ eV}} = 1{,}910 \text{ nm}$$

$$\lambda_{4 \to 2} = \frac{hc}{E_{4 \to 2}} = \frac{(4.14 \times 10^{-15} \text{ eV} \cdot \text{s})(3.00 \times 10^8 \text{ m/s})}{2.55 \text{ eV}} = 487 \text{ nm}$$

$$\lambda_{4 \to 1} = \frac{hc}{E_{4 \to 1}} = \frac{(4.14 \times 10^{-15} \text{ eV} \cdot \text{s})(3.00 \times 10^8 \text{ m/s})}{12.8 \text{ eV}} = 97 \text{ nm}$$

Note that $\lambda_{4 \to 2}$ is in the visible spectrum; this wavelength corresponds to the color blue-green; $\lambda_{4 \to 3}$ is an ultraviolet wavelength, and $\lambda_{4 \to 3}$ is infrared.

WAVE-PARTICLE DUALITY

Light and other electromagnetic waves exhibit wave-like characteristics through interference and diffraction. However, as we saw in the photoelectric effect, light also behaves as if its energy were granular, composed of particles. This is **wave-particle duality**: Electromagnetic radiation propagates like a wave but exchanges energy like a particle.

Since an electromagnetic wave can behave like a particle, can a particle of matter behave like a wave? In 1924, the French physicist Louis de Broglie proposed that the answer is "yes." His conjecture, which has since been supported by experiment, is that a particle of mass m and speed v—and thus, linear momentum p—has an associated wavelength, which is called its **de Broglie wavelength**:

Equation Sheet

$$\lambda = \frac{h}{p}$$

Particles in motion can display wave characteristics and behave as if they had a wavelength.

Since the value of h is so small, ordinary macroscopic objects do not display wave-like behavior. For example, a baseball (mass = 0.15 kg) thrown at a speed of 40 m/s has a de Broglie wavelength of

$$\lambda = \frac{h}{p} = \frac{h}{mv} = \frac{6.63 \times 10^{-34} \text{ J} \cdot \text{s}}{(0.15 \text{ kg})(40 \text{ m/s})} = 1.1 \times 10^{-34} \text{ m}$$

This is much too small to measure. However, with subatomic particles, the wave nature is clearly evident. The 1937 Nobel prize was awarded for experiments that revealed that a stream of electrons exhibited diffraction patterns when scattered by crystals—a behavior that's characteristic of waves.

There's an interesting connection between the de Broglie wavelength for electrons and Bohr's model of quantized orbital radii. Bohr postulated that the electron's orbital angular momentum, $m_e v_n r_n$, had to be an integral multiple of $\hbar = \frac{h}{2\pi}$. The equation

$$m_e v_n r_n = n\frac{h}{2\pi}$$

can be rewritten as follows:

$$2\pi r_n = n\frac{h}{m_e v_n} \quad \Rightarrow \quad 2\pi r_n = n\frac{h}{p_e} \quad \Rightarrow \quad 2\pi r_n = n\lambda_e$$

In this last form, the equation says that the circumference of the electron's orbit must be equal to a whole number of wavelengths in order for it to be stable, a restriction that should remind you of sustained standing waves.

> **Example 4** Electrons in a diffraction experiment are accelerated through a potential difference of 200 V. What is the de Broglie wavelength of these electrons?

Solution. By definition, the kinetic energy of these electrons is 200 eV. Since the relationship between linear momentum and kinetic energy is $p = \sqrt{2mK}$,

$$\lambda = \frac{h}{p} = \frac{h}{\sqrt{2mK}} = \frac{6.63 \times 10^{-34} \text{ J} \cdot \text{s}}{\sqrt{2(9.11 \times 10^{-31} \text{ kg}) \left(200 \text{ eV} \cdot \frac{1.6 \times 10^{-19} \text{ J}}{1 \text{ eV}} \right)}} = 8.7 \times 10^{-11} \text{ m} = 0.087 \text{ nm}$$

This wavelength is characteristic of X-rays.

RELATIVITY

Another important idea in modern physics is the theory of relativity. Relativity has only two postulates:

1. The results of physical experiments will be the same in any non-accelerating reference frames.
2. The speed of light is constant.

For the AP Exam, you are not required to know any of the mathematical framework surrounding relativity, but only have a qualitative understanding of the consequences of the these postulates.

The major consequence of relativity is that time is not absolute, but that space and time are linked together. To see why this must be the case, imagine a car driving down the highway at 65 mph. In non-relativistic physics, the light from the headlight would be moving at the speed of light plus 65 mph. However, the second postulate of relativity states that this cannot be true, and the light from a headlight of a moving car must also be exactly the speed of light.

As the name indicates, relativity describes the consequences of performing experiments in different reference frames and measuring the same results.

When a clock placed on a fast-moving airplane is compared to a clock at rest on the ground, the clock in the airplane shows that less time has passed than the time recorded by the clock on the ground. This phenomenon is known as **time dilation**, and it has been demonstrated experimentally using synchronized atomic clocks.

The speed of light is constant according to the theory of relativity. However, time dilation shows there are disagreements about the amount of time that passes in different reference frames. To be consistent with time dilation, there must also be disagreement about distances. This is known as **length contraction**. A scientist moving along with the object he measures will observe the object to have a different length than a scientist in a different reference frame.

NUCLEAR PHYSICS

The nucleus of the atom is composed of particles called **protons** and **neutrons**, which are collectively called **nucleons**. The number of protons in a given nucleus is called the atom's **atomic number**, and is denoted Z, and the number of neutrons (the **neutron number**) is denoted N. The total number of nucleons, $Z + N$, is called the **mass number** (or **nucleon number**), and is denoted A. The number of protons in the nucleus of an atom defines the element. For example, the element chlorine (abbreviated Cl) is characterized by the fact that the nucleus of every chlorine atom contains 17 protons, so the atomic number of chlorine is 17; but, different chlorine atoms may contain different numbers of neutrons. In fact, about three-fourths of all naturally occurring chlorine atoms have 18 neutrons in their nuclei (mass number = 35), and most of the remaining one-fourth contain 20 neutrons (mass number = 37).

Nuclei that contain the same numbers of protons but different numbers of neutrons are called **isotopes**. The notation for a **nuclide**—the term for a nucleus with specific numbers of protons and neutrons—is to write Z and A before the chemical symbol of the element:

$$_Z^A X$$

The isotopes of chlorine mentioned above would be written as follows:

$$_{17}^{35}Cl \quad \text{and} \quad _{17}^{37}Cl$$

Example 5 How many protons and neutrons are contained in the nuclide $_{29}^{63}Cu$?

Solution. The subscript (the atomic number, Z) gives the number of protons, which is 29. The superscript (the mass number, A) gives the total number of nucleons. Since $A = 63 = Z + N$, we find that $N = 63 - 29 = 34$.

Example 6 The element neon (abbreviated Ne, atomic number 10) has several isotopes. The most abundant isotope contains 10 neutrons, and two others contain 11 and 12. Write symbols for these three nuclides.

Solution. The mass numbers of these isotopes are 10 + 10 = 20, 10 + 11 = 21, and 10 + 12 = 22. So, we'd write them as follows:

$$_{10}^{20}Ne, \quad _{10}^{21}Ne, \quad \text{and} \quad _{10}^{22}Ne$$

Another common notation—which we also use—is to write the mass number after the name of the element. These three isotopes of neon would be written as neon-20, neon-21, and neon-22.

The Nuclear Force

Why wouldn't any nucleus that has more than one proton be unstable? After all, protons are positively charged and would therefore experience a repulsive Coulomb force from each other. Why don't these nuclei explode? And what holds neutrons—which have no electric charge—in the nucleus? These issues are resolved by the presence of another fundamental force, the **strong nuclear force**, which binds neutrons and protons together to form nuclei. Although the strength of the Coulomb force can be expressed by a simple mathematical formula (it's inversely proportional to the square of their separation), the nuclear force is much more complicated; no simple formula can be written for the strength of the nuclear force.

Binding Energy

The masses of the proton and neutron are listed below.

$$\text{proton:} \quad m_p = 1.6726 \times 10^{-27} \text{ kg}$$

$$\text{neutron:} \quad m_n = 1.6749 \times 10^{-27} \text{ kg}$$

Because these masses are so tiny, a much smaller mass unit is used. With the most abundant isotope of carbon (carbon-12) as a reference, the **atomic mass unit** (abbreviated **amu** or simply **u**) is defined as 1/12 the mass of a ^{12}C atom. The conversion between kg and u is 1 u = 1.6605×10^{-27} kg. In terms of atomic mass units,

$$m_p = 1.00728 \text{ u}$$

$$m_n = 1.00867 \text{ u}$$

Now consider the **deuteron**, the nucleus of **deuterium**, an isotope of hydrogen that contains 1 proton and 1 neutron. The mass of a deuteron is 2.01356 u, which is a little *less* than the sum of the individual masses of the proton and neutron. The difference between the mass of any bound nucleus and the sum of the masses of its constituent nucleons is called the **mass defect**, Δm. In the case of the deuteron (symbolized **d**), the mass defect is

$$
\begin{aligned}
m &= (m_p + m_n) - m_d \\
&= (1.00728 \text{ u} + 1.00867 \text{ u}) - (2.01356 \text{ u}) \\
&= 0.00239 \text{ u}
\end{aligned}
$$

What happened to this missing mass? It was converted to energy when the deuteron was formed. In 1905, Einstein gave us the famous equation:

Equation Sheet

$$E = mc^2$$

which tells us how much energy mass contains. Because of this, the mass-difference gives us the amount of energy needed to break the deuteron into a separate proton and neutron. Since this tells us how strongly the nucleus is bound, it is called the **binding energy** of the nucleus.

$$E_B = (\Delta m)c^2$$

Using $E = mc^2$, the energy equivalent of 1 atomic mass unit is

$$E = (1.6605 \times 10^{-27} \text{ kg})(2.9979 \times 10^8 \text{ m / s})^2$$
$$= 1.4924 \times 10^{-10} \text{ J}$$
$$= 1.4924 \times 10^{-10} \text{ J} \cdot \frac{1 \text{ eV}}{1.6022 \times 10^{-19} \text{ J}}$$
$$= 9.31 \times 10^8 \text{ eV}$$
$$= 931 \text{ MeV}$$

In terms of electronvolts, then, the binding energy of the deuteron is

$$E_B \text{ (deuteron)} = 0.00239 \text{ u} \times \frac{931 \text{ MeV}}{1 \text{ u}} = 2.23 \text{ MeV}$$

Since the deuteron contains 2 nucleons, the **binding-energy-per-nucleon** is

$$\frac{2.23 \text{ MeV}}{2 \text{ nucleons}} = 1.12 \text{ MeV/nucleon}$$

This is the lowest value of all nuclides. The highest, 8.8 MeV/nucleon, is for an isotope of nickel, ^{62}Ni. Typically, when nuclei smaller than nickel are fused to form a single nucleus, the binding energy per nucleon increases, which tells us that energy is released in the process. On the other hand, when nuclei *larger* than nickel are *split*, binding energy per nucleon again increases, releasing energy. This is the basis of nuclear fission.

Example 7 What is the maximum wavelength of EM radiation that could be used to photodisintegrate a deuteron?

Solution. The binding energy of the deuteron is 2.23 MeV, so a photon would need to have at least this much energy to break the deuteron into a proton and neutron. Since $E = hf$ and $f = c/\lambda$,

$$E = \frac{hc}{\lambda} \longrightarrow \lambda_{max} = \frac{hc}{E_{min}} = \frac{(4.14 \times 10^{-15} \text{ eV} \cdot \text{s})(3.00 \times 10^8 \text{ m/s})}{2.23 \times 10^6 \text{ eV}} = 5.57 \times 10^{-13} \text{ m}$$

Example 8 The atomic mass of $^{27}_{13}$Al is 26.9815 u. What is its nuclear binding energy per nucleon? (Mass of electron = 0.0005486 u.)

Solution. The nuclear mass of $_{13}^{27}\text{Al}$ is equal to its atomic mass minus the mass of its electrons. Since an aluminum atom has 13 protons, it must also have 13 electrons. So,

$$\text{nuclear mass of } _{13}^{27}\text{Al} = (\text{atomic mass of } _{13}^{27}\text{Al}) - 13m_e$$
$$= 26.9815 \text{ u} - 13(0.0005486 \text{ u})$$
$$= 26.9744 \text{ u}$$

Now, the nucleus contains 13 protons and $27 - 13 = 14$ neutrons, so the total mass of the individual nucleons is

$$M = 13m_p + 14m_n$$
$$= 13(1.00728 \text{ u}) + 14(1.00867 \text{ u})$$
$$= 27.2160 \text{ u}$$

and, the mass defect of the aluminum nucleus is

$$\Delta m = M - m = 27.2160 \text{ u} - 26.9744 \text{ u} = 0.2416 \text{ u}$$

Converting this mass to energy, we can see that

$$E_B = 0.2416 \text{ u} \times \frac{931 \text{ MeV}}{1 \text{ u}} = 225 \text{ MeV}$$

so the binding energy per nucleon is

$$\frac{225 \text{ MeV}}{27} = 8.3 \text{ MeV/nucleon}$$

NUCLEAR REACTIONS

Natural radioactive decay provides one example of a nuclear reaction. Other examples of nuclear reactions include the bombardment of target nuclei with subatomic particles to artificially induce radioactivity, such as the emission of a particle or the splitting of the nucleus (this is **nuclear fission**), and the **nuclear fusion** of small nuclei at extremely high temperatures. In all cases of nuclear reactions that we'll study, nucleon number and charge must be conserved. In order to balance nuclear reactions, we write $_1^1\text{p}$ or $_1^1\text{H}$ for a proton and $_0^1\text{n}$ for a neutron. Gamma-ray photons can also be produced in nuclear reactions; they have no charge or nucleon number and are represented as $_0^0\gamma$.

Matter Cannot be Created nor Destroyed
But it *can* undergo other forms—which does not break this rule. The quickest way to determine whether something undergoes alpha, beta, or gamma decay, is to find the amount of nucleons that have escaped and then to determine which kind of decay happened.

Alpha Decay

When a nucleus undergoes alpha decay, it emits an alpha particle, which consists of two protons and two neutrons and is the same as the nucleus of a helium-4 atom. An alpha particle can be represented as

$$\alpha, \ {}^4_2\alpha, \ \text{or} \ {}^4_2\text{He}$$

Very large nuclei can shed nucleons quickly by emitting one or more alpha particles, for example, radon-222 ($^{222}_{86}\text{Rn}$) is radioactive and undergoes alpha decay.

$$^{222}_{86}\text{Rn} \ \rightarrow \ ^{218}_{84}\text{Po} \ + \ ^4_2\alpha$$

This reaction illustrates two important features of a nuclear reaction.

(1) Mass number is conserved.
(2) Charge is conserved.

The decaying nuclide is known as the **parent**, and the resulting nuclide is known as the **daughter**. (Here, radon-222 is the parent nuclide and polonium-218 is the daughter.) Alpha decay decreases the mass number by 4 and the atomic number by 2. Therefore, alpha decay looks like the following:

$$^A_Z\text{X} \ \rightarrow \ ^{A-4}_{Z-2}\text{X}' \ + \ ^4_2\alpha$$

Beta Decay

There are three subcategories of **beta** (β) decay, called β^-, β^+, and **electron capture (EC)**.

β^- Decay

When the neutron-to-proton ratio is too large, the nucleus undergoes β^- decay, which is the most common form of beta decay. β^- decay occurs when a neutron transforms into a proton and releases an electron. The expelled electron is called a **beta particle**. The transformation of a neutron into a proton and an electron (and another particle, the **electron-antineutrino**, $\bar{\nu}_e$) is caused by the action of the **weak nuclear force**, another of nature's fundamental forces. A common example of a nuclide that undergoes β^- decay is carbon-14, which is used to date archaeological artifacts.

$$^{14}_6\text{C} \ \rightarrow \ ^{14}_7\text{N} \ + \ e^- \ + \ \bar{\nu}_e$$

The reaction is balanced, since $14 = 14 + 0$ and $6 = 7 + (-1)$.

β⁺ Decay

When the neutron-to-proton ratio is too small, the nucleus will undergo β⁺ decay. In this form of beta decay, a proton is transformed into a neutron and a **positron**, $_{+1}^{0}e$ (the electron's **antiparticle**), plus another particle, the **electron-neutrino**, ν_e, which are then both ejected from the nucleus. An example of a positron emitter is fluorine-17.

$$_{9}^{16}\text{F} \rightarrow {}_{8}^{17}\text{O} + e^+ + \nu_e$$

Electron Capture Another way in which a nucleus can increase its neutron-to-proton ratio is to capture an orbiting electron and then cause the transformation of a proton into a neutron. Beryllium-7 undergoes this process.

$$_{4}^{7}\text{Be} + e^- \rightarrow {}_{3}^{7}\text{Li} + \nu_e$$

Gamma Decay

In each of the decay processes defined above, the daughter was a different element than the parent. Radon becomes polonium as a result of α decay, carbon becomes nitrogen as a result of β⁻ decay, fluorine becomes oxygen from β⁺ decay, and beryllium becomes lithium from electron capture. By contrast, gamma decay does not alter the identity of the nucleus; it just allows the nucleus to relax and shed energy. Imagine that potassium-42 undergoes β⁻ decay to form calcium-42.

$$_{19}^{42}\text{K} \rightarrow {}_{20}^{42}\text{Ca}^* + e^- + \bar{\nu}_e$$

The asterisk indicates that the daughter calcium nucleus is left in a high-energy, excited state. For this excited nucleus to drop to its ground state, it must emit a photon of energy, a **gamma ray**, symbolized by γ.

$$_{20}^{42}\text{Ca}^* \rightarrow {}_{20}^{42}\text{Ca} + \gamma$$

Let's sum up the three types of radiation:

Type of Radiation	What Happens?	Charge
Alpha	Particle—a helium nucleus, containing 2 protons and 2 neutrons, is released from the nucleus	Positive 2+
Beta	Particle—a highly energetic electron is released from the nucleus. It has negligible mass. When a beta particle is produced, a neutron divides into a proton and an electron. The electron is ejected from the nucleus. That ejected electron is the beta particle.	Negative 1−
Gamma	Wave—a photon of energy is emitted	No Charge

Example 9 A mercury-198 nucleus is bombarded by a neutron, which causes a nuclear reaction:

$$_{0}^{1}\text{n} + \ _{80}^{198}\text{Hg} \rightarrow \ _{79}^{197}\text{Au} + \ _{?}^{?}\text{X}$$

What's the unknown product particle, X?

Solution. In order to balance the superscripts, we must have $1 + 198 = 197 + A$, so $A = 2$, and the subscripts are balanced if $0 + 80 = 79 + Z$, so $Z = 1$:

$$_{0}^{1}\text{n} + \ _{80}^{198}\text{Hg} \rightarrow \ _{79}^{197}\text{Au} + \ _{1}^{2}\text{X}$$

Therefore, X must be a deuteron, $_{1}^{2}\text{H}$ (or just d).

DISINTEGRATION ENERGY

Nuclear reactions not only produce new nuclei and other subatomic product particles, they also involve the absorption or emission of energy. Nuclear reactions must conserve total energy, so changes in mass are accompanied by changes in energy according to Einstein's equation

$$\Delta E = (\Delta m)c^2$$

A general nuclear reaction is written

$$A + B \rightarrow C + D + Q$$

where Q denotes the **disintegration energy**. If Q is positive, the reaction is **exothermic** (or **exoergic**) and the reaction can occur spontaneously; if Q is negative, the reaction is **endothermic** (or **endoergic**) and the reaction cannot occur spontaneously. The energy Q is calculated as follows:

$$Q = [(m_A + m_B) - (m_C + m_D)]c^2 = (\Delta m)c^2$$

For spontaneous reactions—ones that liberate energy—most of the energy is revealed as kinetic energy of the least massive product nuclei.

Example 10 The process that powers the Sun—and upon which all life on Earth is dependent—is the fusion reaction:

$$4\,_{1}^{1}\text{H} \rightarrow \ _{2}^{4}\alpha + 2\,e^{+} + 2\nu_{e} + \gamma$$

(a) Show that this reaction releases energy.

(b) How much energy is released per proton?

Use the fact that $m_{\alpha} = 4.0015$ u and ignore the mass of the electron-neutrino, ν_{e}.

Solution.

(a)   We need to find the mass difference between the reactants and products:

$$m = 4m_p - (m_\alpha + 2m_e)$$
$$= 4(1.00728 \text{ u}) - [4.0015 \text{ u} + 2(0.0005486 \text{ u})]$$
$$= 0.02652 \text{ u}$$

Since Δm is positive, the reaction is exothermic: Energy is released.

(b)   Converting the mass difference to energy gives

$$Q = 0.02652 \text{ u} \times \frac{931 \text{ MeV}}{1 \text{ u}} = 24.7 \text{ MeV}$$

Since four protons went into the reaction, the energy liberated per proton is

$$\frac{24.7 \text{ MeV}}{4 \text{ p}} = 6.2 \text{ MeV/proton}$$

Example 11 Can the following nuclear reaction occur spontaneously?

$$^4_2\alpha \ + \ ^{14}_7\text{N} \rightarrow ^{17}_8\text{O} + ^1_1\text{H}$$

(The mass of the nitrogen nucleus is 13.9992 u, and the mass of the oxygen nucleus is 16.9947 u.)

Solution. We first figure out the mass equivalent of the disintegration energy:

$$\Delta m = (m_\alpha + m_N) - (m_O + m_p)$$
$$= (4.0015 \text{ u} + 13.9992 \text{ u}) - (16.9947 \text{ u} + 1.00728 \text{ u})$$
$$= -0.00128 \text{ u}$$

Since Δm is negative, this reaction is nonspontaneous; energy must be *supplied* in order for this reaction to proceed. But how much?

$$|Q| = 0.00128 \text{ u} \times \frac{931 \text{ MeV}}{1 \text{ u}} = 1.19 \text{ MeV}$$

Chapter 11 Review Questions

Solutions can be found in Chapter 12.

Section I: Multiple Choice

1. What's the energy of a photon whose wavelength is 2.07 nm ?

   (A) 60 eV
   (B) 600 eV
   (C) 960 eV
   (D) 6000 eV

2. A metal whose work function is 6.0 eV is struck with light of frequency 7.2×10^{15} Hz. What is the maximum kinetic energy of photoelectrons ejected from the metal's surface?

   (A) 7 eV
   (B) 13 eV
   (C) 19 eV
   (D) 24 eV

3. An atom with one electron has an ionization energy of 25 eV. How much energy will be released when the electron makes the transition from an excited energy level, where $E = -16$ eV, to the ground state?

   (A) 9 eV
   (B) 11 eV
   (C) 16 eV
   (D) 25 eV

4. The single electron in an atom has an energy of -40 eV when it's in the ground state, and the first excited state for the electron is at -10 eV. What will happen to this electron if the atom is struck by a stream of photons, each of energy 15 eV ?

   (A) The electron will absorb the energy of one photon and become excited halfway to the first excited state, then quickly return to the ground state, emitting a 15 eV photon in the process.
   (B) The electron will absorb the energy of one photon and become excited halfway to the first excited state, then quickly absorb the energy of another photon to reach the first excited state.
   (C) The electron will absorb two photons and be excited to the first excited state.
   (D) Nothing will happen.

5. What is the de Broglie wavelength of a proton whose linear momentum has a magnitude of 3.3×10^{-23} kg·m/s ?

   (A) 0.0002 nm
   (B) 0.002 nm
   (C) 0.02 nm
   (D) 0.2 nm

6. A partial energy-level diagram for an atom is shown below. What photon energies could this atom emit if it begins in the $n = 3$ state?

   -3 eV _____ $n = 4$

   -5 eV _____ $n = 3$

   -8 eV _____ $n = 2$

   -12 eV _____ $n = 1$ ground state

   (A) 5 eV only
   (B) 3 eV or 7 eV only
   (C) 2 eV, 3 eV, or 7 eV
   (D) 3 eV, 4 eV, or 7 eV

7. The ground-state energy level for He⁺ is -54.4 eV. How much energy must the electron in the ground state of He⁺ absorb in order to be excited to the next higher energy level?

   (A) 13.6 eV
   (B) 27.2 eV
   (C) 40.8 eV
   (D) 68.0 eV

8. What would happen to the energy of a photon if its wavelength were reduced by a factor of 2 ?

   (A) It would decrease by a factor of 4.
   (B) It would decrease by a factor of 2.
   (C) It would increase by a factor of 2.
   (D) It would increase by a factor of 4.

9. In an exothermic nuclear reaction, the difference in mass between the reactants and the products is m, and the energy released is Q. In a separate exothermic nuclear reaction in which the mass difference between reactants and products is $m/4$, how much energy will be released?

(A) $Q/4$
(B) $Q/2$
(C) $(Q/4)c^2$
(D) $(Q/2)c^2$

10. What's the missing particle in the following nuclear reaction?

$$\,^{2}_{1}\text{H} + \,^{63}_{29}\text{Cu} \rightarrow \,^{64}_{30}\text{Zn} + (?)$$

(A) Proton
(B) Neutron
(C) Electron
(D) Positron

11. What's the missing particle in the following nuclear reaction?

$$\,^{196}_{78}\text{Pt} + \,^{1}_{0}\text{n} \rightarrow \,^{197}_{78}\text{Pt} + (?)$$

(A) Proton
(B) Electron
(C) Positron
(D) Gamma

Section II: Free Response

1. The Bohr model of electron energy levels can be applied to any one-electron atom, such as doubly ionized lithium (Li^{2+}). The energy levels for the electron are given by the equation

$$E_n = \frac{Z^2}{n^2}(-13.6 \text{ eV})$$

where Z is the atomic number. The emission spectrum for Li^{2+} contains four spectral lines corresponding to the following wavelengths:

$$11.4 \text{ nm}, \quad 13.5 \text{ nm}, \quad 54.0 \text{ nm}, \quad 72.9 \text{ nm}$$

(a) What's the value of Z for Li^{2+}?

(b) Identify which energy-level transitions give rise to the four wavelengths cited.

(c) Can the emission spectrum for Li^{2+} contain a line corresponding to a wavelength between 54.0 nm and 72.9 nm? If so, calculate its wavelength. If not, briefly explain.

(d) What is the next shortest wavelength in the emission spectrum (closest to, but shorter than, 11.4 nm)?

Summary

○ The energy available to liberate electrons near the surface of a metal (the photoelectric effect) is proportional to the frequency of the incident photon. This idea is expressed in $E = hf$ where h is Planck's constant. ($h = 6.63 \times 10^{-34}$ J·s).

○ The work function (ϕ) indicates the amount of energy needed to liberate the electron. There is a minimum frequency needed to liberate the electrons: $f_0 = \dfrac{\phi}{h}$. The kinetic energy of the emitted electron is $K_{max} = hf - \phi$.

○ Particles in motion have wavelike properties: $\lambda = \dfrac{h}{p}$ where p is the momentum of the particle. Combining this with $E = hf$ and $c = f\lambda$ yields $E = pc$.

○ Standard notation for an element is $_{Z}^{A}X$, where A is the mass number, Z is the number of protons in the nucleus, and $A = Z+N$, where N is the number of neutrons in the nucleus.

○ A nuclear reaction produces new nuclei, other subatomic particles, and the absorption or emission of energy. The change in mass between the reactants and the products tells how much energy is released (exothermic or $+Q$) or how much energy is needed to produce the reaction (endothermic or $-Q$) in the general equation $A + B \rightarrow C + D + Q$, where A and B are reactants and C and D are products. This energy released or absorbed is given by $\Delta E = (\Delta m)c^2$.

Chapter 12
Solutions to
Chapter Review
Questions

CHAPTER 3 REVIEW QUESTIONS

Section I: Multiple Choice

1. **D** Since Point X is 5 m below the surface of the water, the pressure due to the water at X, P_X, is $\rho g h_X = \rho g (5 \text{ m})$, where ρ is the density of water. Because Point Y is 4 m below the surface of the water, the pressure due to the water at Y, p_Y, is $\rho g (4 \text{ m})$. Therefore,

$$\frac{P_X}{P_Y} = \frac{\rho g (5 \text{ m})}{\rho g (4 \text{ m})} = \frac{5}{4} \quad \Rightarrow \quad 4P_X = 5P_Y$$

2. **C** Because the top of the box is at a depth of $D - z$ below the surface of the liquid, the pressure on the top of the box is

$$P = P_{atm} + \rho g h = P_{atm} + \rho g (D - z)$$

The area of the top of the box is $A = xy$, so the force on the top of the box is

$$F = PA = [P_{atm} + \rho g (D - z)]xy$$

3. **C** Let P_0 be the pressure of the gas at the surface of the liquid. Then the pressures at points Y and Z are, respectively,

$$P_Y = P_0 + \rho g (1 \text{ m})$$
$$P_Z = P_0 + \rho g (3 \text{ m})$$

Subtracting the first equation from the second, we get $P_Z - P_Y = \rho g (2 \text{ m})$. Because we're given the values of P_Y and P_Z, we know that $P_Z - P_Y = 16{,}000$ Pa, so we can write

$$16{,}000 \text{ Pa} = \rho g (2 \text{ m}) \quad \Rightarrow \quad \rho g = 8{,}000 \, \tfrac{\text{Pa}}{\text{m}}$$

Substituting this into the equation for P_Y, we find that

$$P_Y = P_0 + \rho g (1 \text{ m}) \quad \Rightarrow \quad P_0 = P_Y - \rho g (1 \text{ m}) = (13{,}000 \text{ Pa}) - (8{,}000 \, \tfrac{\text{Pa}}{\text{m}})(1 \text{ m}) = 5{,}000 \text{ Pa}$$

4. **A** The density of the plastic cube is

$$\rho = \frac{m}{V} = \frac{m}{l^3} = \frac{100 \text{ kg}}{(\frac{1}{2} \text{ m})^3} = 800 \, \tfrac{\text{kg}}{\text{m}^3}$$

This is 4/5 the density of the water, so 4/5 of the cube's volume is submerged; this means that 1/5 of the cube's volume is above the surface of the water.

5. **B** The buoyant force on the styrofoam block is $F_{buoy} = \rho_L V g$, and the weight of the block is $F_g = m_S g = \rho_S V g$. Because $\rho_L > \rho_S$, the net force on the block is upward and has magnitude

$$F_{net} = F_{buoy} - F_g = (\rho_L - \rho_S)Vg$$

Therefore, by Newton's Second Law, we have

$$a = \frac{F_{net}}{m} = \frac{(\rho_L - \rho_S)Vg}{\rho_S V} = \left(\frac{\rho_L}{\rho_S} - 1\right)g$$

6. **B** The buoyant force acting on the ball is

$$F_{buoy} = \rho_{water} V_{sub} g = \rho_{water} V g = (1000 \tfrac{kg}{m^3})(5 \times 10^{-3} \text{ m}^3)(10 \tfrac{N}{kg}) = 50 \text{ N}$$

The weight of the ball is

$$F_g = \rho_{ball} V g = (400 \tfrac{kg}{m^3})(5 \times 10^{-3} \text{ m}^3)(10 \tfrac{N}{kg}) = 20 \text{ N}$$

Because the upward force on the ball, F_{buoy}, balances the total downward force, $F_g + F_T$, the tension in the string is

$$F_T = F_{buoy} - F_g = 50 \text{ N} - 20 \text{ N} = 30 \text{ N}$$

7. **A** If the object weighs 100 N less when completely submerged in water, the buoyant force must be

$$F_{buoy} = \rho_{water} V_{sub} g = \rho_{water} V g = 100 \text{ N} \quad \Rightarrow \quad V = \frac{100 \text{ N}}{\rho_{water} g} = \frac{100 \text{ N}}{(1000 \tfrac{kg}{m^3})(10 \tfrac{N}{kg})} = 10^{-2} \text{ m}^3$$

Now that we know the volume of the object, we can figure out its weight:

$$F_g = mg = \rho_{object} V g = (2000 \tfrac{kg}{m^3})(10^{-2} \text{ m}^3)(10 \tfrac{N}{kg}) = 200 \text{ N}$$

8. **A** The cross-sectional diameter at Y is 3 times the cross-sectional diameter at X, so the cross-sectional area at Y is $3^2 = 9$ times that at X. The Continuity Equation tells us that the flow speed, v, is inversely proportional to the cross-sectional area, A. So, if A is 9 times greater at Point Y than it is at X, then the flow speed at Y is 1/9 the flow speed at X; that is, $v_Y = (1/9)v_X = (1/9)(6 \text{ m/s}) = 2/3 \text{ m/s}$.

9. **D** Each side of the rectangle at the bottom of the conduit is 1/4 the length of the corresponding side at the top. Therefore, the cross-sectional area at the bottom is $(1/4)^2 = 1/16$ the cross-sectional area at the top. The Continuity Equation tells us that the flow speed, v, is inversely proportional to the cross-sectional area, A. So, if A at the bottom is 1/16 the value of A at the top, then the flow speed at the bottom is 16 times the flow speed at the top.

10. **D** We'll apply Bernoulli's Equation to a point at the pump (Point 1) and at the nozzle (the exit point, Point 2). We'll choose the level of Point 1 as the horizontal reference level; this makes $y_1 = 0$ and $y_2 = 1$ m. Now, because the cross-sectional diameter decreases by a factor of 10 between Points 1 and 2, the cross-sectional area decreases by a factor of $10^2 = 100$, so flow speed must increase by a factor of 100; that is, $v_2 = 100v_1 = 100(0.4 \text{ m/s}) = 40$ m/s. Because Point 2 is exposed to the air, the pressure there is P_{atm}. Bernoulli's Equation becomes

$$P_1 + \frac{1}{2}\rho v_1^2 = P_{atm} + \rho g y_2 + \frac{1}{2}\rho v_2^2$$

Therefore,

$$
\begin{aligned}
P_1 - P_{atm} &= \rho g y_2 + \frac{1}{2}\rho v_2^2 - \frac{1}{2}\rho v_1^2 \\
&= (1{,}000\,\tfrac{\text{kg}}{\text{m}^3})(10\text{m/s}^2)(1\text{ m}) + \frac{1}{2}(1{,}000\,\tfrac{\text{kg}}{\text{m}^3})(40\text{ m/s})^2 - \frac{1}{2}(1{,}000\,\tfrac{\text{kg}}{\text{m}^3})(0.4\text{ m/s})^2 \\
&\approx (1{,}000\,\tfrac{\text{kg}}{\text{m}^3})(10\text{m/s}^2)(1\text{ m}) + \frac{1}{2}(1{,}000\,\tfrac{\text{kg}}{\text{m}^3})(40\text{ m/s})^2 \\
&= (10{,}000\text{ Pa}) + (800{,}000\text{ Pa}) \\
&= 810{,}000\text{ Pa} \\
&= 810\text{ kPa}
\end{aligned}
$$

Section II: Free Response

1. (a) The pressure at the top surface of the block is $P_{top} = P_{atm} + \rho_L g h$. Since the area of the top of the block is $A = xy$, the force on the top of the block has magnitude

$$F_{top} = P_{bottom} A = \left(P_{atm} + \rho_L g h\right)xy$$

The pressure at the bottom of the block is $P_{bottom} = P_{atm} + \rho_L g(h + z)$. Since the area of the bottom face of the block is also $A = xy$, the force on the bottom surface of the block has magnitude

$$F_{bottom} = P_{bottom} A = \left[P_{atm} + \rho_L g(h + z)\right]xy$$

These forces are sketched below:

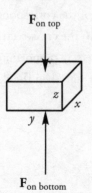

(b) Each of the other four faces of the block (left and right, front and back) is at an average depth of $h + \frac{1}{2}z$, so the average pressure on each of these four sides is

$$\overline{P}_{\text{sides}} = P_{\text{atm}} + \rho_L g\left(h + \frac{1}{2}z\right)$$

The left and right faces each have area $A = xz$, so the magnitude of the average force on this pair of faces is

$$\overline{F}_{\text{left and right}} = \overline{P}_{\text{sides}}A = \left[P_{\text{atm}} + \rho_L g\left(h + \frac{1}{2}z\right)\right]xz$$

The front and back faces each have area $A = yz$, so the magnitude of the average force on this pair of faces is

$$\overline{F}_{\text{front and back}} = \overline{P}_{\text{sides}}A = \left[P_{\text{atm}} + \rho_L g\left(h + \frac{1}{2}z\right)\right]yz$$

These forces are sketched below:

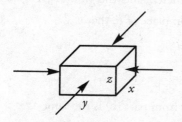

(c) The four forces sketched in part (b) add up to zero, so the total force on the block due to the pressure is the sum of $F_{\text{on top}}$ and $F_{\text{on bottom}}$; because $F_{\text{on bottom}} > F_{\text{on top}}$, this total force points upward and its magnitude is

$$F_{\text{on bottom}} - F_{\text{on top}} = \left[P_{\text{atm}} + \rho_L g(h + z)\right]xy - (P_{\text{atm}} + \rho_L gh)xy = \rho_L gxyz$$

(d) By Archimedes' Principle, the buoyant force on the block is upward and has magnitude

$$F_{\text{buoy}} = \rho_L V_{\text{sub}}g = \rho_L Vg = \rho_L xyzg$$

This is the same as the result we found in part (c).

(e) The weight of the block is

$$F_g = mg = \rho_B Vg = \rho_B xyzg$$

If F_T is the tension in the string, then the total upward force on the block, $F_T + F_{\text{buoy}}$, must balance the downward force, F_g; that is, $F_T + F_{\text{buoy}} = F_g$, so

$$F_T = F_g - F_{\text{buoy}} = \rho_B xyzg - \rho_L xyzg = xyzg(\rho_B - \rho_L)$$

2. (a) See the section on Torricelli's Theorem for the derivation of the efflux speed from the hole; applying Bernoulli's Equation to a point on the surface of the water in the tank (Point 1) and a point at the hole (Point 2), the assumption that $v_1 \approx 0$ leads to the result

$$v = \sqrt{2gh}$$

(b) The initial velocity of the water, as it emerges from the hole, is horizontal. Since there's no initial vertical velocity, the time t required to drop the distance $y = D - h$ to the ground is found as follows:

$$y = \frac{1}{2}gt^2 \quad \Rightarrow \quad D - h = \frac{1}{2}gt^2 \quad \Rightarrow \quad t = \sqrt{\frac{2(D-h)}{g}}$$

Therefore, the horizontal distance the water travels is

$$x = v_x t = \sqrt{2gh} \cdot \sqrt{\frac{2(D-h)}{g}} = 2\sqrt{h(D-h)}$$

(c) The second hole would be at a depth of $h/2$ below the surface of the water, so the horizontal distance it travels—from the edge of the tank to the point where it hits the ground—is given by the same formula we found in part (b) except it will have $h/2$ in place of h; that is,

$$x_2 = 2\sqrt{\frac{1}{2}h(D - \frac{1}{2}h)}$$

If both streams land at the same point, then the value of x from part (b) is the same as x_2:

$$2\sqrt{h(D-h)} = 2\sqrt{\frac{1}{2}h(D - \frac{1}{2}h)}$$
$$h(D-h) = \frac{1}{2}h(D - \frac{1}{2}h)$$
$$D - h = \frac{1}{2}(D - \frac{1}{2}h)$$
$$4D - 4h = 2D - h$$
$$-3h = -2D$$
$$h = \frac{2}{3}D$$

(d) Once again, we apply Bernoulli's Equation to a point on the surface of the water in the tank (Point 1) and to a point at the hole (Point 2). We'll choose the ground level as our horizontal reference level; then $y_1 = D$ and $y_2 = D - h$. If v_1 is the flow speed of Point 1—that is, the speed with which the water level in the tank drops—and v_2 is the efflux speed from the hole, then, by the Continuity Equation, $A_1 v_1 = A_2 v_2$, where A_1 and A_2 are the cross-sectional areas at Points 1 and 2, respectively. Therefore, $v_1 = (A_2/A_1)v_2$. Bernoulli's Equation then becomes

$$P_1 + \rho g D + \frac{1}{2}\rho v_1^2 = P_2 + \rho g (D-h) + \frac{1}{2}\rho v_2^2$$

Since $P_1 = P_2 = P_{atm}$, these terms cancel out; and substituting $v_1 = (A_2/A_1)v_2$, we have

$$\rho g D + \frac{1}{2}\rho\left(\frac{A_2}{A_1}v_2\right)^2 = \rho g(D - h) + \frac{1}{2}\rho v_2^2$$

$$\frac{1}{2}\rho v_2^2\left[\left(\frac{A_2}{A_1}\right)^2 - 1\right] = -\rho g h$$

$$v_2 = \sqrt{\frac{2gh}{1 - \left(\dfrac{A_2}{A_1}\right)^2}}$$

Now, since $A_1 = \pi R^2$ and $A_2 = \pi r^2$, this final equation can be written as

$$v_2 = \sqrt{\frac{2gh}{1 - \left(\dfrac{r}{R}\right)^4}}$$

[Note that if $r \ll R$, then $(r/R)^4 \approx 0$, and the equation above reduces to $v_2 = \sqrt{2gh}$, as in part (a).]

3. (a) Point X is at a depth of h_1 below Point 1, where the pressure is P_1. Therefore, the hydrostatic pressure at X is $P_X = P_1 + \rho_F g h_1$.

   (b) Point Y is at a depth of $h_2 + d$ below Point 2, where the pressure is P_2. The column of static fluid above Point Y contains fluid of density of depth h_2 and fluid of density ρ_F of depth d. Therefore, the hydrostatic pressure at Y is $P_Y = P_2 + \rho_F g h_2 + \rho_V g d$.

   (c) First, notice that Points 1 and 2 are at the same horizontal level; therefore, the heights y_1 and y_2 are the same, and the terms $\rho_F g y_1$ and $\rho_F g y_2$ will cancel out of the equation. Bernoulli's Equation then becomes

   $$P_1 + \frac{1}{2}\rho_F v_1^2 = P_2 + \frac{1}{2}\rho_F v_2^2$$

   By the Continuity Equation, we have $A_1 v_1 = A_2 v_2$, so $v_1 = (A_2/A_1)v_2$. Therefore,

   $$P_1 - P_2 = \frac{1}{2}\rho_F v_2^2 - \frac{1}{2}\rho_F v_1^2$$

   $$= \frac{1}{2}\rho_F v_2^2 - \frac{1}{2}\rho_F\left(\frac{A_2}{A_1}v_2\right)^2$$

   $$= \frac{1}{2}\rho_F v_2^2\left[1 - \left(\frac{A_2}{A_1}\right)^2\right]$$

(d)   In parts (a) and (b) above, we found that $P_X = P_1 + \rho_F g h_1$ and $P_Y = P_2 + \rho_F g h_2 + \rho_v g d$. Since $P_X = P_Y$, we have

$$P_1 + \rho_F g h_1 = P_2 + \rho_F g h_2 + \rho_V g d$$

so

$$\begin{aligned} P_1 - P_2 &= \rho_F g (h_2 - h_1) + \rho_V g d \\ &= \rho_F g(-d) + \rho_V g d \\ &= (\rho_V - \rho_F) g d \end{aligned}$$

(e)   In parts (c) and (d), we found two expressions for $P_1 - P_2$. Setting them equal to each other gives

$$\frac{1}{2} \rho_F v_2^2 \left[1 - \left(\frac{A_2}{A_1} \right)^2 \right] = (\rho_V - \rho_F) g d$$

$$v_2^2 = \frac{\rho_V - \rho_F}{\rho_F} \cdot \frac{2gd}{1 - \left(\dfrac{A_2}{A_1} \right)^2}$$

$$v_2 = \sqrt{ \frac{2gd \left(\dfrac{\rho_V}{\rho_F} - 1 \right)}{1 - \left(\dfrac{A_2}{A_1} \right)^2} }$$

The flow rate in the pipe is

$$f = A_2 v_2 = A_2 \sqrt{ \frac{2gd \left(\dfrac{\rho_V}{\rho_F} - 1 \right)}{1 - \left(\dfrac{A_2}{A_1} \right)^2} }$$

Since

$$f = A_2 \sqrt{ \frac{2g \left(\dfrac{\rho_V}{\rho_F} - 1 \right)}{1 - \left(\dfrac{A_2}{A_1} \right)^2} } \cdot \sqrt{d}$$

we see that f is proportional to $\sqrt{d}$, as desired.

CHAPTER 4 REVIEW QUESTIONS

Section I: Multiple Choice

1. **B** Because the average kinetic energy of a molecule of gas is directly proportional to the temperature of the sample, the fact that the gases are at the same temperature—since they're in the same container at thermal equilibrium—tells us that the molecules have the same average kinetic energy. The ratio of their kinetic energies is therefore equal to 1.

2. **D** From the Ideal Gas Law, we know that $P = nRT/V$. If both T and V are doubled, the ratio T/V remains unchanged, so P remains unchanged.

3. **A** The work done on the gas during a thermodynamic process is equal to the area of the region in the P–V diagram above the V-axis and below the path the system takes from its initial state to its final state. Since the area below path 1 is the greatest, the work done on the gas during the transformation along path 1 is the greatest.

4. **C** During an isothermal change, ΔU is always zero.

5. **B** Because the gas is confined, n remains constant, and because we're told the volume is fixed, V remains constant as well. Since R is a universal constant, the Ideal Gas Law, $PV = nRT$, tells us that P and T are proportional. Therefore, if T increases by a factor of 2, then so does P.

6. **D** Neither (A) nor (B) can be correct. Using $PV = nRT$, both containers have the same V, n is the same, P is the same, and R is a universal constant. Therefore, T must be the same for both samples. The kinetic theory of gases predicts that the rms speed of the gas molecules in a sample of molar mass M and temperature T is

$$v_{\text{rms}} = \sqrt{\frac{3RT}{M}}$$

    Hydrogen has a smaller molar mass than does helium, so v_{rms} for hydrogen must be greater than v_{rms} for helium (because both samples are at the same T).

7. **A** By convention, work done *on* the gas sample is designated as positive, so in the First Law of Thermodynamics, $\Delta U = Q + W$, we must write $W = +320$ J. Therefore, $Q = \Delta U - W = 560$ J $- 320$ J $= +240$ J. Positive Q denotes heat *in*.

8. **C** No work is done during the step from state a to state b because the volume doesn't change. Therefore, the work done from a to c is equal to the work done from b to c. Since the pressure remains constant (this step is isobaric), we find that

$$W = -P\Delta V = -(3.0 \times 10^5 \text{ Pa})[(10 - 25) \times 10^{-3} \text{ m}^3] = 4{,}500 \text{ J}$$

9. **C** Choice (A) is wrong because "no heat exchanged between the gas and its surroundings" is the definition of *adiabatic*, not *isothermal*. Choice (B) cannot be correct since the step described in the question is isothermal; by definition, the temperature does not change. This also eliminates (D) and supports (C). If the sample could be brought back to its initial state *and* have a 100% conversion of heat to work, *that* would violate the Second Law. The Second Law states that heat cannot be completely converted to work with no other change taking place. In this case, there are changes taking place: The pressure decreases and the volume increases.

10. **C** The second law of thermodynamics indicates that energy will flow from a hot object to a cool object, making the final temperature of the hot object cooler and the cold object warmer.

Section II: Free Response

1. (a) First, let's calculate ΔU_{acb}. Using path *acb*, the question tells us that $Q = +70$ J and $W = -30$ J (W is negative here because it is the *system* that does the work). The First Law, $\Delta U = Q + W$, tells us that $\Delta U_{acb} = +40$ J. Because $\Delta U_{a \to b}$ does not depend on the path taken from a to b, we must have $\Delta U_{ab} = +40$ J, and $\Delta U_{ba} = -\Delta U_{ab} = -40$ J. Thus, -40 J $= Q_{ba} + W_{ba}$, where Q_{ba} and W_{ba} are the values along the curved path from b to a. Since $Q_{ba} = -60$ J, it follows that $W_{ba} = +20$ J. Therefore, the surroundings do 20 J of work on the system.

    (b) Again, using the fact that $\Delta U_{a \to b}$ does not depend on the path taken from a to b, we know that $\Delta U_{adb} = +40$ J, as computed above. Writing $\Delta U_{adb} = QW_{adb} + W_{adb}$, if $W_{adb} = -10$ J, it follows that $Q_{adb} = +50$ J. That is, the system absorbs 50 J of heat.

    (c) For the process *db*, there is no change in volume, so $W_{db} = 0$. Therefore, $\Delta U_{db} = Q_{db} + W_{db} = Q_{db}$. Now, since $U_{ab} = +40$ J, the fact that $U_a = 0$ J implies that $U_b = 40$ J, so $\Delta U_{db} = U_b - U_d = 40$ J $- 30$ J $= 10$ J. Thus, $Q_{db} = 10$ J. Now let's consider the process *ad*. Since $W_{adb} = W_{ad} + W_{db} = W_{ad} + 0 = W_{ad}$, the fact that $W_{adb} = -10$ J [computed in part (c)] tells us that $W_{ad} = -10$ J. Because $\Delta U_{ad} = U_d - U_a = 30$ J, it follows from $\Delta U_{ad} = Q_{ad} + W_{ad}$ that $Q_{ad} = \Delta U_{ad} - W_{ad} = 30$ J $- (-10$ J$) = 40$ J.

    (d) The process *adbca* is cyclic, so ΔU is zero. Because this cyclic process is traversed *counterclockwise* in the P–V diagram, we know that W is *positive*. Then, since $\Delta U = Q + W$, it follows that Q must be negative.

2. (a) (i) Use the Ideal Gas Law:

$$T_a = \frac{P_a V_a}{nR} = \frac{(2.4 \times 10^5 \text{ Pa})(12 \times 10^{-3} \text{ m}^3)}{(0.4 \text{ mol})(8.31 \text{ J/mol} \cdot \text{K})} = 870 \text{ K}$$

(ii) Since state b is on the isotherm with state a, the temperature of state b must also be 870 K.

(iii) Use the Ideal Gas Law:

$$T_c = \frac{P_c V_c}{nR} = \frac{(0.6 \times 10^5 \text{ Pa})(12 \times 10^{-3} \text{ m}^3)}{(0.4 \text{ mol})(8.31 \text{ J/mol} \cdot \text{K})} = 220 \text{ K}$$

(b) (i) Since step ab takes place along an isotherm, the temperature of the gas does not change, so neither does the internal energy; $\Delta U_{ab} = 0$.

(ii) By the First Law of Thermodynamics, $\Delta U_{bc} = Q_{bc} + W_{bc}$. Since step bc is isobaric (constant pressure), we have

$$Q_{bc} = nC_P \Delta T_{bc} = (0.4 \text{ mol})(29.1 \text{ J/mol} \cdot \text{K})(220 \text{ K} - 870 \text{ K}) = -7,600 \text{ J}$$

Next,

$$W_{bc} = -P \Delta V_{bc} = -(0.6 \times 10^5 \text{ Pa})[(12 - 48) \times 10^{-3} \text{ m}^3] = 2,200 \text{ J}$$

Therefore,

$$\Delta U_{bc} = Q_{bc} + W_{bc} = (-7,600 \text{ J}) + 2,200 \text{ J} = -5,400 \text{ J}$$

(iii) By the First Law of Thermodynamics, $\Delta U_{ca} = Q_{ca} + W_{ca}$. Since step ca is isochoric (constant volume), we have

$$Q_{ca} = nC_V \Delta T_{ca} = (0.4 \text{ mol})(20.8 \text{ J/mol} \cdot \text{K})(870 \text{ K} - 220 \text{ K}) = 5,400 \text{ J}$$

and $W_{ca} = 0$. Therefore, $\Delta U_{ca} = Q_{ca} + W_{ca} = (5,400 \text{ J}) + 0 = 5,400 \text{ J}$.

Note that part (iii) could have been answered as follows:

$$\Delta U_{aa} = \Delta U_{ab} + \Delta U_{bc} + \Delta U_{ca}$$
$$0 = 0 + \Delta U_{bc} + \Delta U_{ca}$$
$$\Delta U_{ca} = -\Delta U_{bc}$$
$$= -(-5,400 \text{ J})$$
$$= 5,400 \text{ J}$$

(c) Using the equation given, we find that

$$W_{ab} = -nRT \cdot \ln\frac{V_b}{V_a} = -(0.4 \text{ mol})(8.31 \text{ J/mol} \cdot \text{K})(870 \text{ K}) \cdot \ln\frac{48 \times 10^{-3} \text{ m}^3}{12 \times 10^{-3} \text{ m}^3}$$
$$= -4,000 \text{ J}$$

(d) The total work done over the cycle is equal to the sum of the values of the work done over each step:

$$
\begin{aligned}
W_{cycle} &= W_{ab} + W_{bc} + W_{ca} \\
&= W_{ab} + W_{bc} \\
&= (-4{,}000 \text{ J}) + (2{,}200 \text{ J}) \\
&= -1{,}800 \text{ J}
\end{aligned}
$$

(e) (i) and (ii) By the First Law of Thermodynamics,

$$
\begin{aligned}
\Delta U_{ab} &= Q_{ab} + W_{ab} \\
0 &= Q_{ab} + W_{ab} \\
Q_{ab} &= -W_{ab} \\
&= 4{,}000 \text{ J} \quad \text{[from part (c)]}
\end{aligned}
$$

Since Q is positive, this represents heat *absorbed* by the gas.

(f) The maximum possible efficiency is the efficiency of a Carnot engine:

$$
e_C = 1 - \frac{T_C}{T_H} = 1 - \frac{220 \text{ K}}{870 \text{ K}} = 75\%
$$

CHAPTER 5 REVIEW QUESTIONS

Section I: Multiple Choice

1. **D** Electrostatic force obeys an inverse-square law: $F_E \propto 1/r^2$. Therefore, if r increases by a factor of 3, then F_E decreases by a factor of $3^2 = 9$.

2. **C** The strength of the electric force is given by kq^2/r^2, and the strength of the gravitational force is Gm^2/r^2. Since both of these quantities have r^2 in the denominator, we simply need to compare the numerical values of kq^2 and Gm^2. There's no contest: Since

$$
kq^2 = (9 \times 10^9 \text{ N·m}^2/\text{C}^2)(1 \text{ C})^2 = 9 \times 10^9 \text{ N·m}^2
$$

and

$$
Gm^2 = (6.7 \times 10^{-11} \text{ N·m}^2/\text{kg}^2)(1 \text{ kg})^2 = 6.7 \times 10^{-11} \text{ N·m}^2
$$

we see that $kq^2 \gg Gm^2$, so F_E is much stronger than F_G.

3. **C** If the net electric force on the center charge is zero, the electrical repulsion by the $+2q$ charge must balance the electrical repulsion by the $+3q$ charge:

$$\frac{1}{4\pi\varepsilon_0}\frac{(2q)(q)}{x^2} = \frac{1}{4\pi\varepsilon_0}\frac{(3q)(q)}{y^2} \;\Rightarrow\; \frac{2}{x^2} = \frac{3}{y^2} \;\Rightarrow\; \frac{y^2}{x^2} = \frac{3}{2} \;\Rightarrow\; \frac{y}{x} = \sqrt{\frac{3}{2}}$$

4. **D** Since P is equidistant from the two charges, and the magnitudes of the charges are identical, the strength of the electric field at P due to $+Q$ is the same as the strength of the electric field at P due to $-Q$. The electric field vector at P due to $+Q$ points away from $+Q$, and the electric field vector at P due to $-Q$ points toward $-Q$. Since these vectors point in the same direction, the net electric field at P is (E to the right) + (E to the right) = ($2E$ to the right).

5. **C** The acceleration of the small sphere is

$$a = \frac{F_E}{m} = \frac{1}{4\pi\varepsilon_0}\frac{Qq}{mr^2}$$

As r increases (that is, as the small sphere is pushed away), a decreases. However, since a is always positive, the small sphere's speed, v, is always increasing.

6. **B** Since $\mathbf{F}_E$ (on q) = $q\mathbf{E}$, it must be true that $\mathbf{F}_E$ (on $-2q$) = $-2q\mathbf{E} = -2\mathbf{F}_E$.

7. **C** All excess electric charge on a conductor resides on the outer surface.

Section II: Free Response

1. (a) From the figure below, we have $F_{1\text{-}2} = F_1/\cos 45°$.

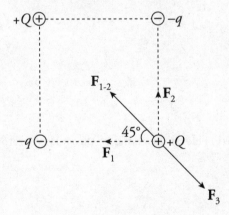

Since the net force on $+Q$ is zero, we want $F_{1-2} = F_3$. If s is the length of each side of the square, then:

$$F_{1-2} = F_3 \implies \frac{F_1}{\cos 45°} = F_3 \implies \frac{1}{\cos 45°} \frac{1}{4\pi\varepsilon_0} \frac{Qq}{s^2} = \frac{1}{4\pi\varepsilon_0} \frac{Q^2}{(s\sqrt{2})^2}$$

$$\sqrt{2} \cdot q = \frac{Q}{2}$$

Given that $\dfrac{1}{\cos 45}$ is the square root of 2 and cancelling like terms yields: $q = \dfrac{Q}{2\sqrt{2}}$

(b)　No. If $q = Q/2\sqrt{2}$, as found in part (a), then the net force on $-q$ is not zero.

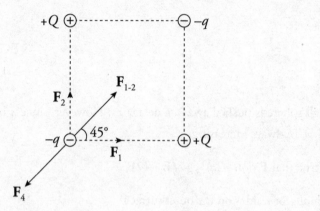

This is because F_{1-2} is greater than F_4, resulting in a net force toward the center of the square, as the following calculations show:

$$F_{1-2} = \frac{F_1}{\cos 45°} = \sqrt{2} \frac{1}{4\pi\varepsilon_0} \frac{Qq}{s^2} = \sqrt{2} \frac{1}{4\pi\varepsilon_0} \frac{Q\frac{Q}{2\sqrt{2}}}{s^2} = \frac{1}{8\pi\varepsilon_0} \frac{Q^2}{s^2}$$

but

$$F_4 = \frac{1}{4\pi\varepsilon_0} \frac{q^2}{(s\sqrt{2})^2} = \frac{1}{4\pi\varepsilon_0} \frac{\left(\frac{Q}{2\sqrt{2}}\right)^2}{(s\sqrt{2})^2} = \frac{1}{64\pi\varepsilon_0} \frac{Q^2}{s^2}$$

(c)　By symmetry, $E_1 = E_2$ and $E_3 = E_4$, so the net electric field at the center of the square is zero:

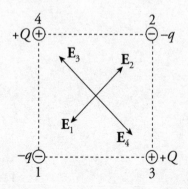

2. (a) The magnitude of the electric force on Charge 1 is

$$F_1 = \frac{1}{4\pi\varepsilon_0}\frac{(Q)(2Q)}{(a+2a)^2} = \frac{1}{18\pi\varepsilon_0}\frac{Q^2}{a^2}$$

The direction of $\mathbf{F}_1$ is directly away from Charge 2; that is, in the $+y$ direction, so

$$\mathbf{F}_1 = \frac{1}{18\pi\varepsilon_0}\frac{Q^2}{a^2}\,\hat{\mathbf{j}}$$

(b) The electric field vectors at the origin due to Charge 1 and due to Charge 2 are

$$\mathbf{E}_1 = \frac{1}{4\pi\varepsilon_0}\frac{Q}{a^2}(-\hat{\mathbf{j}}) \quad \text{and} \quad \mathbf{E}_2 = \frac{1}{4\pi\varepsilon_0}\frac{2Q}{(2a)^2}(+\hat{\mathbf{j}}) = \frac{1}{8\pi\varepsilon_0}\frac{Q}{a^2}(+\hat{\mathbf{j}})$$

Therefore, the net electric field at the origin is

$$\mathbf{E} = \mathbf{E}_1 + \mathbf{E}_2 = \frac{1}{4\pi\varepsilon_0}\frac{Q}{a^2}(-\hat{\mathbf{j}}) + \frac{1}{8\pi\varepsilon_0}\frac{Q}{a^2}(+\hat{\mathbf{j}}) = \frac{1}{8\pi\varepsilon_0}\frac{Q}{a^2}(-\hat{\mathbf{j}})$$

(c) No. The only point on the x-axis where the individual electric field vectors due to each of the two charges point in exactly opposite directions is the origin (0, 0). But at that point, the two vectors are not equal and thus do not cancel.

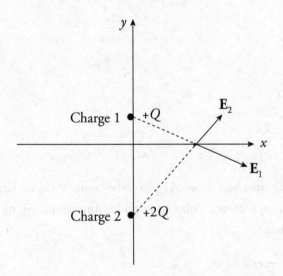

Therefore, at no point on the x-axis could the total electric field be zero.

(d)   Yes. There will be a Point P on the y-axis between the two charges,

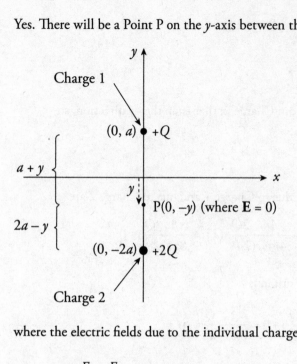

where the electric fields due to the individual charges will cancel each other out.

$$E_1 = E_2$$

$$\frac{1}{4\pi\varepsilon_0}\frac{Q}{(a+y)^2} = \frac{1}{4\pi\varepsilon_0}\frac{2Q}{(2a-y)^2}$$

$$\frac{1}{(a+y)^2} = \frac{2}{(2a-y)^2}$$

$$(2a-y)^2 = 2(a+y)^2$$

$$0 = y^2 + 8ay - 2a^2$$

$$y = \frac{-8a \pm \sqrt{(8a)^2 - 4(-2a^2)}}{2}$$

$$= (-4 \pm 3\sqrt{2})a$$

Disregarding the value $y = (-4 - 3\sqrt{2})a$ (because it would place the point P below Charge 2 on the y-axis, where the electric field vectors do not point in opposite directions), we have that $\mathbf{E} = \mathbf{0}$ at the point $P = (0, -y) = (0, (4 - 3\sqrt{2})a)$.

(e)   Use the result of part (b) with Newton's Second Law:

$$\mathbf{a} = \frac{\mathbf{F}}{m} = \frac{-q\mathbf{E}}{m} = \frac{-q}{m}\left[\frac{1}{8\pi\varepsilon_0}\frac{Q}{a^2}(-\hat{\mathbf{j}})\right] = \frac{1}{8\pi\varepsilon_0}\frac{qQ}{ma^2}(+\hat{\mathbf{j}})$$

CHAPTER 6 REVIEW QUESTIONS

Section I: Multiple Choice

1. **C**   By definition, $\Delta U_E = -W_E$, so if W_E is negative, then ΔU_E is positive. This implies that the potential energy, U_E, increases.

2. **B**   The work required to assemble the configuration is equal to the potential energy of the system:

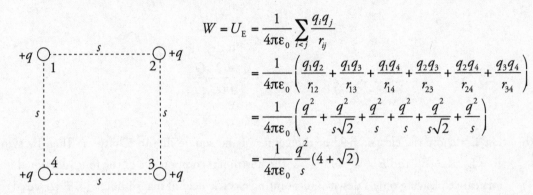

$$W = U_E = \frac{1}{4\pi\varepsilon_0} \sum_{i<j} \frac{q_i q_j}{r_{ij}}$$

$$= \frac{1}{4\pi\varepsilon_0}\left(\frac{q_1 q_2}{r_{12}} + \frac{q_1 q_3}{r_{13}} + \frac{q_1 q_4}{r_{14}} + \frac{q_2 q_3}{r_{23}} + \frac{q_2 q_4}{r_{24}} + \frac{q_3 q_4}{r_{34}} \right)$$

$$= \frac{1}{4\pi\varepsilon_0}\left(\frac{q^2}{s} + \frac{q^2}{s\sqrt{2}} + \frac{q^2}{s} + \frac{q^2}{s} + \frac{q^2}{s\sqrt{2}} + \frac{q^2}{s} \right)$$

$$= \frac{1}{4\pi\varepsilon_0} \frac{q^2}{s}(4 + \sqrt{2})$$

3. **B**   Use the definition $\Delta V = -W_E/q$. If an electric field accelerates a negative charge doing positive work on it, then $W_E > 0$. If $q < 0$, then $-W_E/q$ is positive. Therefore, ΔV is positive, which implies that V increases.

4. **D**   By definition,

    $V_{A\to B} = \Delta U_E/q$, so $V_B - V_A = \Delta U_E/q$

5. **B**   Because **E** is uniform, the potential varies linearly with distance from either plate ($\Delta V = Ed$). Since Points 2 and 4 are at the same distance from the plates, they lie on the same equipotential. (The equipotentials in this case are planes parallel to the capacitor plates.)

6. **B**   By definition, $W_E = -q\Delta V$, which gives

    $W_E = -q(V_B - V_A) = -(-0.05 \text{ C})(100 \text{ V} - 200 \text{ V}) = -5 \text{ J}$

    Note that neither the length of the segment AB nor that of the curved path from A to B is relevant.

Section II: Free Response

1.   (a) Labeling the four charges as given in the diagram, we get

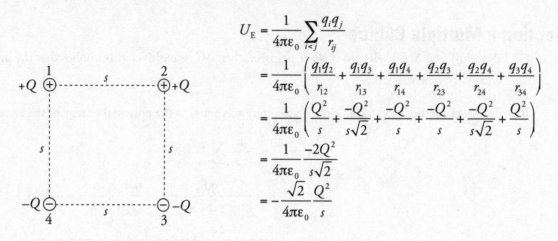

$$U_E = \frac{1}{4\pi\varepsilon_0} \sum_{i<j} \frac{q_i q_j}{r_{ij}}$$

$$= \frac{1}{4\pi\varepsilon_0}\left(\frac{q_1 q_2}{r_{12}} + \frac{q_1 q_3}{r_{13}} + \frac{q_1 q_4}{r_{14}} + \frac{q_2 q_3}{r_{23}} + \frac{q_2 q_4}{r_{24}} + \frac{q_3 q_4}{r_{34}}\right)$$

$$= \frac{1}{4\pi\varepsilon_0}\left(\frac{Q^2}{s} + \frac{-Q^2}{s\sqrt{2}} + \frac{-Q^2}{s} + \frac{-Q^2}{s} + \frac{-Q^2}{s\sqrt{2}} + \frac{Q^2}{s}\right)$$

$$= \frac{1}{4\pi\varepsilon_0}\frac{-2Q^2}{s\sqrt{2}}$$

$$= -\frac{\sqrt{2}}{4\pi\varepsilon_0}\frac{Q^2}{s}$$

  (b) Let $\mathbf{E}_i$ denote the electric field at the center of the square due to Charge i. Then by symmetry, $\mathbf{E}_1 = \mathbf{E}_3$, $\mathbf{E}_2 = \mathbf{E}_4$, and $E_1 = E_2 = E_3 = E_4$. The horizontal components of the four individual field vectors cancel, leaving only a downward-pointing electric field of magnitude $E_{total} = 4E_1 \cos 45°$:

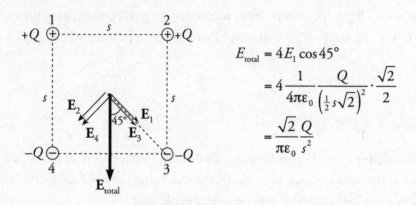

$$E_{total} = 4E_1 \cos 45°$$

$$= 4\frac{1}{4\pi\varepsilon_0}\frac{Q}{\left(\frac{1}{2}s\sqrt{2}\right)^2}\cdot\frac{\sqrt{2}}{2}$$

$$= \frac{\sqrt{2}}{\pi\varepsilon_0}\frac{Q}{s^2}$$

  (c) The potential at the center of the square is zero:

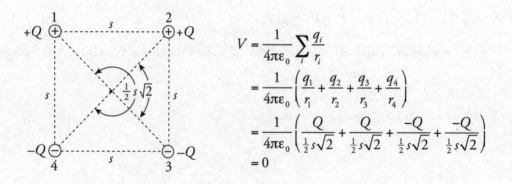

$$V = \frac{1}{4\pi\varepsilon_0}\sum_i \frac{q_i}{r_i}$$

$$= \frac{1}{4\pi\varepsilon_0}\left(\frac{q_1}{r_1} + \frac{q_2}{r_2} + \frac{q_3}{r_3} + \frac{q_4}{r_4}\right)$$

$$= \frac{1}{4\pi\varepsilon_0}\left(\frac{Q}{\frac{1}{2}s\sqrt{2}} + \frac{Q}{\frac{1}{2}s\sqrt{2}} + \frac{-Q}{\frac{1}{2}s\sqrt{2}} + \frac{-Q}{\frac{1}{2}s\sqrt{2}}\right)$$

$$= 0$$

(d)   At every point on the center horizontal line shown, $r_1 = r_4$ and $r_2 = r_3$, so V will equal zero (just as it does at the center of the square):

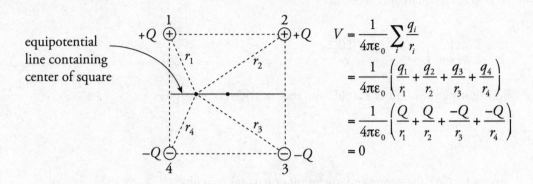

$$V = \frac{1}{4\pi\varepsilon_0} \sum_i \frac{q_i}{r_i}$$

$$= \frac{1}{4\pi\varepsilon_0} \left(\frac{q_1}{r_1} + \frac{q_2}{r_2} + \frac{q_3}{r_3} + \frac{q_4}{r_4} \right)$$

$$= \frac{1}{4\pi\varepsilon_0} \left(\frac{Q}{r_1} + \frac{Q}{r_2} + \frac{-Q}{r_3} + \frac{-Q}{r_4} \right)$$

$$= 0$$

(e)   The work done by the electric force as q is displaced from Point A to Point B is given by the equation $W_E = -q\Delta V_{A \to B} = -q(V_B - V_A) = -qV_B$ (since $V_A = 0$).

$$W_E = -qV_B = -q\left[\frac{1}{4\pi\varepsilon_0} \sum_i \frac{q_i}{r_{i \to B}} \right]$$

$$= \frac{-q}{4\pi\varepsilon_0} \left(\frac{q_1}{r_{1 \to B}} + \frac{q_2}{r_{2 \to B}} + \frac{q_3}{r_{3 \to B}} + \frac{q_4}{r_{4 \to B}} \right)$$

$$= \frac{-q}{4\pi\varepsilon_0} \left(\frac{Q}{\frac{1}{2}s} + \frac{Q}{\frac{1}{2}s} + \frac{-Q}{\sqrt{s^2 + (\frac{1}{2}s)^2}} + \frac{-Q}{\sqrt{s^2 + (\frac{1}{2}s)^2}} \right)$$

$$= \frac{-q}{4\pi\varepsilon_0} \left(\frac{4Q}{s} + \frac{-2Q}{\frac{\sqrt{5}}{2}s} \right)$$

$$= \frac{-qQ}{\pi\varepsilon_0 s} \left(1 - \frac{1}{\sqrt{5}} \right)$$

2.   (a)   The capacitance is $C = \varepsilon_0 A/d$. Since the plates are rectangular, the area A is equal to Lw, so $C = \varepsilon_0 Lw/d$.

(b) and (c) Since the electron is attracted upward, the top plate must be the positive plate:

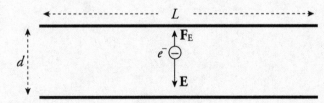

(d)   The acceleration of the electron is $a = F_E/m = qE/m = eE/m$, vertically upward. Therefore, applying Big Five #3 for vertical motion, $\Delta y = v_{0y}t + \frac{1}{2}a_y t^2$, we get

$$\Delta y = \frac{1}{2}\frac{eE}{m}t^2 \quad (1)$$

To find t, notice that $v_{0x} = v_0$ remains constant (because there is no horizontal acceleration). Therefore, the time necessary for the electron to travel the horizontal distance L is $t = L/v_0$. In this time, Δy is $d/2$, so Equation (1) becomes

$$\frac{d}{2} = \frac{1}{2}\frac{eE}{m}\left(\frac{L}{v_0}\right)^2 \quad \Rightarrow \quad E = \frac{dmv_0^2}{eL^2}$$

(e) Substituting the result of part (d) into the equation $\Delta V = Ed$ gives

$$\Delta V = \frac{d^2mv_0^2}{eL^2}$$

Since $Q = C\Delta V$ (by definition), the result of part (a) now gives

$$Q = \frac{\varepsilon_0 Lw}{d} \cdot \frac{d^2mv_0^2}{eL^2} = \frac{\varepsilon_0 wdmv_0^2}{eL}$$

(f) Applying the equation $U_E = \frac{1}{2}C(\Delta V)^2$, we get

$$U_E = \frac{1}{2} \cdot \frac{\varepsilon_0 Lw}{d} \cdot \left(\frac{d^2mv_0^2}{eL^2}\right)^2 = \frac{\varepsilon_0 wd^3m^2v_0^4}{2e^2L^3}$$

3. (a) Outside the sphere, the sphere behaves as if all the charge were concentrated at the center. Inside the sphere, the electrostatic field is zero:

$$E(r) = \begin{cases} 0 & (r < a) \\ \dfrac{1}{4\pi\varepsilon_0}\dfrac{Q}{r^2} & (r > a) \end{cases}$$

(b) On the surface and outside the sphere, the electric potential is $(\frac{1}{4}\pi\varepsilon_0)(Q/r)$. Within the sphere, V is constant (because $E = 0$) and equal to the value on the surface. Therefore,

$$V(r) = \begin{cases} \dfrac{1}{4\pi\varepsilon_0}\dfrac{Q}{a} & (r \le a) \\ \dfrac{1}{4\pi\varepsilon_0}\dfrac{Q}{r} & (r > a) \end{cases}$$

(c) See diagrams:

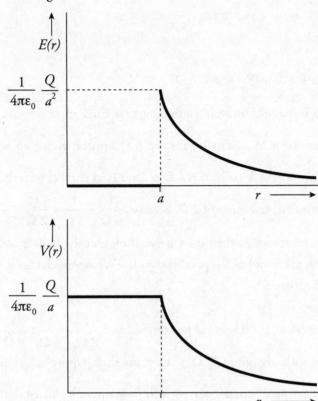

CHAPTER 7 REVIEW QUESTIONS

Section I: Multiple Choice

1. **A** Let ρ_S denote the resistivity of silver and let A_S denote the cross-sectional area of the silver wire. Then

$$R_B = \frac{\rho_B L}{A_B} = \frac{(5\rho_S)L}{4^2 A_S} = \frac{5}{16}\frac{\rho_S L}{A_S} = \frac{5}{16}R_S$$

2. **C** The equation $I = V/R$ implies that increasing V by a factor of 2 will cause I to increase by a factor of 2.

3. **C** Use the equation $P = V^2/R$:

$$P = \frac{V^2}{R} \quad \Rightarrow \quad R = \frac{V^2}{P} = \frac{(120 \text{ V})^2}{60 \text{ W}} = 240 \text{ }\Omega$$

4. **B** The current through the circuit is

$$I = \frac{\varepsilon}{r+R} = \frac{40 \text{ V}}{(5 \text{ }\Omega)+(15 \text{ }\Omega)} = 2 \text{ A}$$

Therefore, the voltage drop across R is $V = IR = (2 \text{ A})(15 \text{ }\Omega) = 30 \text{ V}$.

5. **D** The 12 Ω and 4 Ω resistors are in parallel and are equivalent to a single 3 Ω resistor, because $\frac{1}{12 \text{ }\Omega} + \frac{1}{4 \text{ }\Omega} = \frac{1}{3 \text{ }\Omega}$. This 3 Ω resistor is in series with the top 3 Ω resistor, giving an equivalent resistance in the top branch of $3 + 3 = 6$ Ω. Finally, this 6 Ω resistor is in parallel with the bottom 3 Ω resistor, giving an overall equivalent resistance of 2 Ω, because $\frac{1}{6 \text{ }\Omega} + \frac{1}{3 \text{ }\Omega} = \frac{1}{2 \text{ }\Omega}$.

6. **C** If each of the identical bulbs has resistance R, then the current through each bulb is ε/R. This is unchanged if the middle branch is taken out of the parallel circuit. (What *will* change is the total amount of current provided by the battery.)

7. **B** The three parallel resistors are equivalent to a single 2 Ω resistor, because $\frac{1}{8 \text{ }\Omega} + \frac{1}{4 \text{ }\Omega} + \frac{1}{8 \text{ }\Omega} = \frac{1}{2 \text{ }\Omega}$. This 2 Ω resistance is in series with the given 2 Ω resistor, so their equivalent resistance is 2 Ω + 2 Ω = 4 Ω. Therefore, three times as much current will flow through this equivalent 4 Ω resistance in the top branch as through the parallel 12 Ω resistor in the bottom branch, which implies that the current through the bottom branch is 3 A, and the current through the top branch is 9 A. The voltage drop across the 12 Ω resistor is therefore $V = IR = (3 \text{ A})(12 \text{ }\Omega) = 36 \text{ V}$.

8. **D** Since points a and b are grounded, they're at the same potential (call it zero).

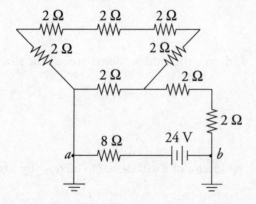

Traveling from b to a across the battery, the potential increases by 24 V, so it must decrease by 24 V across the 8 Ω resistor as we reach point a. Thus, $I = V/R = (24 \text{ V})/(8 \text{ }\Omega) = 3 \text{ A}$.

9. **D** The equation $P = I^2R$ gives

$$P = (0.5 \text{ A})^2(100 \text{ }\Omega) = 25 \text{ W} = 25 \text{ J/s}$$

Therefore, in 20 s, the energy dissipated as heat is

$$E = Pt = (25 \text{ J/s})(20 \text{ s}) = 500 \text{ J}$$

10. **D**

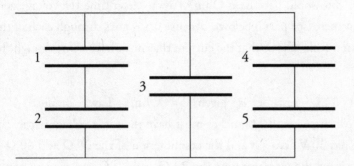

Capacitors 1 and 2 are in series, so their equivalent capacitance is $C_{1-2} = C/2$. (This is obtained from the equation $1/C_{1-2} = 1/C_1 + 1/C_2 = 1/C + 1/C = 2/C$.) Capacitors 4 and 5 are also in series, so their equivalent capacitance is $C_{4-5} = C/2$. The capacitances C_{1-2}, C_3, and C_{4-5} are in parallel, so the overall equivalent capacitance is $(C/2) + C + (C/2) = 2C$.

Section II: Free Response

1. (a) The two parallel branches, the one containing the 40 Ω resistor and the other a total of 120 Ω, is equivalent to a single 30 Ω resistance. This 30 Ω resistance is in series with the three 10 Ω resistors, giving an overall equivalent circuit resistance of 10 Ω + 10 Ω + 30 Ω + 10 Ω = 60 Ω. Therefore, the current supplied by the battery is $I = V/R = (120\text{V})/(60) = 2$ A, so it must supply energy at a rate of $P = IV = (2 \text{ A})(120 \text{ V}) = 240 \text{ W}$.

    (b) Since three times as much current will flow through the 40 Ω resistor as through the branch containing 120 Ω of resistance, the current through the 40 Ω resistor must be 1.5 A.

    (c) (i) $V_a - V_b = IR_{20} + IR_{100} = (0.5 \text{ A})(20 \text{ }\Omega) + (0.5 \text{ A})(100 \text{ }\Omega) = 60 \text{ V}$.

    (ii) Point a is at the higher potential (current flows from high to low potential).

    (d) Because energy is equal to power multiplied by time, we get

$$E = Pt = I^2Rt = (0.5 \text{ A})^2(100 \text{ }\Omega)(10 \text{ s}) = 250 \text{ J}$$

(e) Using the equation $R = \rho L/A$, with $A = \pi r^2$, we find

$$R = \frac{\rho L}{\pi r^2} \quad \Rightarrow \quad r = \sqrt{\frac{\rho L}{\pi R}} = \sqrt{\frac{(0.45\ \Omega \cdot m)(0.04\ m)}{\pi(100\ \Omega)}} = 0.0076\ m = 7.6\ mm$$

2. (a) There are many ways to solve this problem. If you notice that each of the three branches of the parallel section is 60 Ω, then they must all have the same current flowing through them. The currents through the 20 Ω, 40 Ω, and other 60 Ω resistors are all 0.5 A.

   If you had not noticed this, you would have used Ohm's Law to determine the voltage across the resistors and proceeded from there (see part b below). Because 0.5 A goes through each of the three pathways, Kirchhoff's Junction Rule tells us that the current that must have come though the 10 Ω resistor is 1.5 A.

   (b) The voltage across the 60 Ω resistor is given by Ohm's Law. Because $V = IR$, $V = (0.5\ A)(60\ \Omega) = 30\ V$. All three parallel branches must have the same voltage across them, so the other 60 Ω resistor also has 30 V across it and the combination of the 20 Ω and 60 Ω resistor must also be 30 V. To determine the voltage across the 20 Ω and 40 Ω resistor you can rely on the previously solved currents of 0.5A and Ohm's Law to yield $V = IR = (0.5\ A)(20\ \Omega) = 10\ V$ and $V = (0.5\ A)(40\ \Omega) = 20\ V$.

   You also could have used the ratio of the resistors. That is, we know the two voltages must sum to 30 V and the voltage drop across the 40 Ω must be twice the amount across the 20 Ω. The voltage across the 10 Ω resistor can be found using Ohm's Law: $V = (1.5\ A)(10\ \Omega) = 15\ V$.

   (c) The equivalent resistance of the circuit can be solved either by adding the resistances or by using Ohm's Law.

   If you want to add resistances, start by summing the 20 Ω and 40 Ω resistors to get 60 Ω. Then add the three parallel branches using $\frac{1}{R_P} = \sum_i \frac{1}{R_i}$ or $\frac{1}{R_P} = \sum \frac{1}{60\ \Omega} + \frac{1}{60\ \Omega} + \frac{1}{60\ \Omega}$, which becomes 20 Ω. Then adding this section in series to the 10 Ω resistor to get $R_p = R_1 + R_2 = 10\ \Omega + 20\ \Omega = 30\ \Omega$.

   You could have also realized that the total voltage drop across the battery is 45 V (15 V across the 10 Ω resistor and 30 V across the parallel branch). Using Ohm's Law again $R_{eq} = \frac{V_B}{I_B}$ or $R_{eq} = \frac{45V}{1.5A} = 30\ \Omega$.

CHAPTER 8 REVIEW QUESTIONS

Section I: Multiple Choice

1. **C, D** Choice (A) is false. The magnetic field lines due to a current-carrying wire encircle the wire in closed loops. Choice (B) is false, since the magnetic force is always perpendicular to the charged particle's velocity vector, it cannot do work on the charged particle; therefore, it cannot change the particle's kinetic energy. Choice (C) is true. If the charged particle's velocity is parallel (or anti-parallel) to the magnetic field lines, then the particle will feel no magnetic force. Choice (D) is also true. Magnetic fields are created by moving charges, $F = qvB \sin \theta$.

2. **C** The magnitude of the magnetic force is $F_B = qvB$, so the acceleration of the particle has magnitude

$$a = \frac{F_B}{m} = \frac{qvB}{m} = \frac{(4.0 \times 10^{-9} \text{ C})(3 \times 10^4 \text{ m/s})(0.1 \text{ T})}{2 \times 10^{-4} \text{ kg}} = 0.06 \text{ m/s}^2$$

3. **B** By the right-hand rule, the direction of $\mathbf{v} \times \mathbf{B}$ is into the plane of the page. Since the particle carries a negative charge, the magnetic force it feels will be out of the page.

4. **C** Since $\mathbf{F}_B$ is always perpendicular to $\mathbf{v}$, $\mathbf{v}$ cannot be upward or downward in the plane of the page; this eliminates (A) and (B). Because the charge is positive, the direction of $\mathbf{F}_B$ will be the same as the direction of $\mathbf{v} \times \mathbf{B}$. In order for $\mathbf{v} \times \mathbf{B}$ to be downward in the plane of the page, the right-hand rule implies that $\mathbf{v}$ must be out of the plane of the page.

5. **A** The magnetic force provides the centripetal force on the charged particle. Therefore,

$$qvB = \frac{mv^2}{r} \quad \Rightarrow \quad qB = \frac{mv}{r} \quad \Rightarrow \quad mv = qBr \quad \Rightarrow \quad p = qBr$$

6. **C** The strength of the magnetic field at a distance r from a long, straight wire carrying a current I is given by the equation $B = (\mu_0/2\pi)(I/r)$. Therefore,

$$\frac{\mu_0}{2\pi} \frac{I}{r} = \frac{(4\pi \times 10^{-7} \text{ T} \cdot \text{m/A})}{2\pi} \frac{10 \text{ A}}{0.02 \text{ m}} = 1 \times 10^{-4} \text{ T}$$

7. **D** By Newton's Third Law, neither (A) nor (B) can be correct. Also, as we learned in Chapter 8, Example 9, if two parallel wires carry current in the same direction, the magnetic force between them is attractive; this eliminates (C). Therefore, the answer must be (D). The strength of the magnetic field at a distance r from a long, straight wire carrying a current I_1 is given by the equation $B_1 = (\mu_0/2\pi)(I_1/r)$. The magnetic force on a wire of length ℓ carrying a current I through a magnetic field $\mathbf{B}$ is $I(\ell \times \mathbf{B})$, so the force on Wire #2 (F_{B2}) due to the magnetic field of Wire #1 (B_1) is

$$F_{B2} = I_2 \ell B_1 = I_2 \ell \frac{\mu_0}{2\pi} \frac{I_1}{r}$$

which implies

$$\frac{F_{B2}}{\ell} = \frac{\mu_0}{2\pi} \frac{I_1 I_2}{r} = \frac{(4\pi \times 10^{-7} \text{ N/A}^2)}{2\pi} \frac{(5 \text{ A})(10 \text{ A})}{0.01 \text{ m}} = 0.001 \text{ N/m}$$

8. **D** The strength of the magnetic field at a distance r from a long, straight wire carrying a current I is given by the equation $B = (\mu_0/2\pi)(I/r)$. Therefore, the strength of the magnetic field at Point P due to either wire is $B = (\mu_0/2\pi)(I/\frac{1}{2}d)$. By the right-hand rule, the direction of the magnetic field at P due to the top wire is into the plane of the page and the direction of the magnetic field at P due to the bottom wire is out of the plane of the page. Since the two magnetic field vectors at P have the same magnitude and opposite directions, the net magnetic field at Point P is zero.

9. **C** Use the right-hand rule for wires. If you point your thumb to the right and wrap your fingers along the wire, you will note that the magnetic field goes into the page when you are below the wire and comes out of the page above the wire. This allows us to eliminate (A) and (B). Because $B = \frac{\mu_o I}{2\pi r}$, the closer we are to the wire the stronger the magnetic field. Choice (C) is closer, so it is the correct answer.

10. **C** Magnetic fields point from north to south. Therefore, the magnetic field between the two magnets is toward the right of the page. Use the right-hand rule. Because the B field is to the right and the charges through the wire flow to the bottom of the page, the force must be out of the page.

Section II: Free Response

1. (a) The acceleration of an ion of charge q is equal to F_E/m. The electric force is equal to qE, where $E = V/d$. Therefore, $a = qV/(dm)$.

   (b) Using $a = qV/(dm)$ and the equation $v^2 = v_0^2 + 2ad = 2ad$, we get

   $$v^2 = 2\frac{qV}{dm}d \quad \Rightarrow \quad v = \sqrt{\frac{2qV}{m}}$$

   As an alternate solution, notice that the change in the electrical potential energy of the ion from the source S to the entrance to the magnetic-field region is equal to qV; this is equal to the gain in the particle's kinetic energy.

   Therefore,

   $$qV = \frac{1}{2}mv^2 \quad \Rightarrow \quad v = \sqrt{\frac{2qV}{m}}$$

(c)   (i) and (ii) Use the right-hand rule. Since **v** points to the right and **B** is into the plane of the page, the direction of **v** × **B** is upward. Therefore, the magnetic force on a positively charged particle (cation) will be upward, and the magnetic force on a negatively charged particle (anion) will be downward. The magnetic force provides the centripetal force that causes the ion to travel in a circular path. Therefore, a cation would follow Path 1 and an anion would follow Path 2.

(d)   Since the magnetic force on the ion provides the centripetal force,

$$qvB = \frac{mv^2}{r} \quad \Rightarrow \quad qvB = \frac{mv^2}{\frac{1}{2}y} \quad \Rightarrow \quad m = \frac{qBy}{2v}$$

Now, by the result of part (b),

$$m = \frac{qBy}{2\sqrt{\dfrac{2qV}{m}}} \quad \Rightarrow \quad m^2 = \frac{q^2 B^2 y^2}{\dfrac{8qV}{m}} \quad \Rightarrow \quad m^2 = \frac{mq^2 B^2 y^2}{8qV} \quad \Rightarrow \quad m = \frac{qB^2 y^2}{8V}$$

(e)   Since the magnetic force cannot change the speed of a charged particle, the time required for the ion to hit the photographic plate is equal to the distance traveled (the length of the semicircle) divided by the speed computed in part (b):

$$t = \frac{s}{v} = \frac{\pi \cdot \dfrac{1}{2}y}{\sqrt{\dfrac{2qV}{m}}} = \frac{1}{2}\pi y \sqrt{\frac{m}{2qV}}$$

(f)   Since the magnetic force **F**$_B$ is always perpendicular to a charged particle's velocity vector **v**, it can do no work on the particle. Thus, the answer is zero.

2. (a) Because a photon has no charge, it will experience no force and therefore travel in a straight line.

(b) From $r = \dfrac{mv}{qB}$, we get $r = \dfrac{(2.23 \times 10^{-27}\,\text{kg})(40.1\,\text{m/s})}{(1.07 \times 10^{-19}\,C)(6.0 \times 10^{-8}\,T)} = 13.9$ m.

(c) From $r = \dfrac{mv}{qB}$, we get $q = \dfrac{mv}{Br}$. So $q = \dfrac{(7.49 \times 10^{-27}\,\text{kg})(41.5\,\text{m/s})}{(6.0 \times 10^{-8}\,\text{T})(92.7\,\text{m})} = 5.58 \times 10^{-20}$ C. To determine the sign of the charge, use the right-hand rule. Because the magnetic field is into the page and the charge spins in a clockwise manner, it would not obey the right-hand rule that works for all positive charges. Therefore, the charge must be negative.

CHAPTER 9 REVIEW QUESTIONS

Section I: Multiple Choice

1. **D** Since **v** is upward and **B** is out of the page, the direction of **v** × **B** is to the right. Therefore, free electrons in the wire will be pushed to the left, leaving an excess of positive charge at the right. Therefore, the potential at Point *b* will be higher than at Point *a*, by $\mathcal{E} = vBL$ (motional emf).

2. **A** As we discussed in this chapter, the magnitude of the emf induced between the ends of the rod is $\mathcal{E} = BLv = (0.5\text{ T})(0.2\text{ m})(3\text{ m/s}) = 0.3$ V. Since the resistance is 10 Ω, the current induced will be $I = V/R = (0.3\text{ V})/(10\text{ Ω}) = 0.03$ A. To determine the direction of the current, we can note that since positive charges in the rod are moving to the left and the magnetic field points into the plane of the page, the right-hand rule tells us that the magnetic force, $q\mathbf{v} \times \mathbf{B}$, points downward. Since the resulting force on the positive charges in the rod is downward, so is the direction of the induced current.

3. **A** The magnetic field through the loop is $B = \mu_0 nI$. Since its area is $A = \pi r^2$, the magnetic flux through the loop is $\Phi_B = BA = (\mu_0 nI)(\pi r^2)$. If the current changes (with $\Delta I/\Delta t = -a$), then the magnetic flux through the loop changes, which, by Faraday's Law, implies that an emf (and a current) will be induced. We get

$$\mathcal{E} = -\frac{\Delta \Phi_B}{\Delta t} = -\frac{\Delta(\mu_0 nI \cdot \pi r^2)}{\Delta t} = -(\mu_0 n\pi r^2)\frac{\Delta I}{\Delta t} = -\mu_0 \pi nr^2(-a) = \mu_0 \pi nr^2 a$$

Since the magnetic flux into the page is decreasing, the direction of the induced current will be clockwise (opposing a *decreasing into-the-page flux* means that the induced current will create more into-the-page flux).

4. **C** By definition, magnetic field lines emerge from the north pole and enter at the south pole. Therefore, as the north pole is moved upward through the loop, the upward magnetic flux increases. To oppose an increasing upward flux, the direction of the induced current will be clockwise (as seen

from above) to generate some downward magnetic flux. Now, as the south pole moves away from the center of the loop, there is a decreasing upward magnetic flux, so the direction of the induced current will be counterclockwise.

5. **D**   Since the current in the straight wire is steady, there is no change in the magnetic field, no change in magnetic flux, and, therefore, no induced emf or current.

Section II: Free Response

1. (a) $\varepsilon = B\ell v \Rightarrow (2 \text{ T})(0.4 \text{ m})(1 \text{ m/s})$ or 0.8V.

    (b) $V = IR$ becomes $I = \dfrac{V}{R}$ or $I = \dfrac{0.8V}{20\Omega} = 0.04$ A. The direction is given by Lenz's Law. Current flows in order to oppose the change in the magnetic flux. Because there is suddenly a new flux "in," current flows to produce an outward flux. This would be in the counterclockwise direction.

    (c) $\varepsilon = B\ell v \Rightarrow (2 \text{ T})(0.2 \text{ m})(1 \text{ m/s})$ or 0.4 V

    (d)   In order to get the 0.8 V you would need $v = \dfrac{\varepsilon}{B\ell} \Rightarrow \dfrac{0.8 \text{ V}}{(2 \text{ T})(0.2 \text{ m})}$ or $v = 2.0$ m/s. You also could have noticed that if you cut the length in half, you have to compensate by doubling the speed.

CHAPTER 10 REVIEW QUESTIONS

Section I: Multiple Choice

1. **B**   From the equation $\lambda f = c$, we find that

    $$\lambda = \frac{c}{f} = \frac{3.0 \times 10^8 \text{ m/s}}{1.0 \times 10^{18} \text{ Hz}} = 3.0 \times 10^{-10} \text{ m}$$

2. **C**   Since the fringe is bright, the waves must interfere constructively. This implies that the difference in path lengths must be a whole number times the wavelength, eliminating (A) and (B). The central maximum is equidistant from the two slits, so $\Delta \ell = 0$ there. At the first bright fringe above the central maximum, we have $\Delta \ell = \lambda$.

3.   **C**   First, eliminate (A) and (B): The index of refraction is never smaller than 1. Refer to the following diagram:

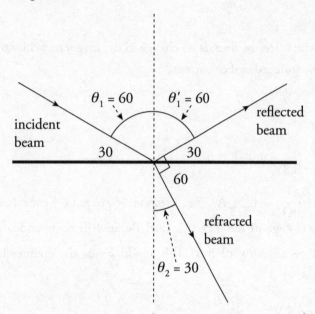

Since the reflected and refracted beams are perpendicular to each other, we have $\theta_2 = 30°$. Snell's Law then becomes

$$n_1 \sin \theta_1 = n_2 \sin \theta_2$$
$$1 \cdot \sin 60° = n_2 \sin 30°$$
$$\frac{1}{2}\sqrt{3} = n_2 \cdot \frac{1}{2}$$
$$\sqrt{3} = n_2$$

4.   **A**   The frequency is unchanged, but because the speed of light in diamond is less than in air, the wavelength of the light in diamond is shorter than its wavelength in air:

$$\lambda_{\text{in diamond}} = \frac{1}{f} v_{\text{in diamond}} = \frac{1}{f} \frac{c}{n_{\text{diamond}}} = \frac{1}{n_{\text{diamond}}} \lambda_{\text{in air}} = \frac{500 \text{ nm}}{2.5} = 200 \text{ nm}$$

5.   **A**   If the speed of light is less in Medium 2 than in Medium 1, then Medium 2 must have the higher index of refraction; that is, $n_2 > n_1$. Snell's Law then implies that $\theta_2 < \theta_1$: The beam will refract toward the normal upon transmission into Medium 2.

6.   **A**   The critical angle for total internal reflection is computed as follows:

$$n_1 \sin \theta_c = n_2 \sin 90°$$
$$\sin \theta_c = \frac{n_2}{n_1} = \frac{1.45}{2.90} = \frac{1}{2} \quad \Rightarrow \quad \theta_c = 30°$$

Total internal reflection can happen only if the incident beam originates in the medium with the higher index of refraction and strikes the interface of the other medium at an angle of incidence greater than the critical angle.

7. **D** If $s_o = 60$ cm and $f = 40$ cm, the mirror equation tells us that

$$\frac{1}{s_o} + \frac{1}{s_i} = \frac{1}{f} \quad \Rightarrow \quad \frac{1}{60 \text{ cm}} + \frac{1}{s_i} = \frac{1}{40 \text{ cm}} \quad \Rightarrow \quad \frac{1}{s_i} = \frac{1}{120 \text{ cm}} \quad \Rightarrow \quad s_i = 120 \text{ cm}$$

And, since s_i is positive, the image is real.

8. **D** Because the image is virtual, we must write the image distance, s_i, as a negative quantity: $s_i = -20$ cm. Now, with $s_o = 60$ cm, the mirror equation gives

$$\frac{1}{s_o} + \frac{1}{s_i} = \frac{1}{f} \quad \Rightarrow \quad \frac{1}{60 \text{ cm}} + \frac{1}{-20 \text{ cm}} = \frac{1}{f} \quad \Rightarrow \quad \frac{1}{f} = \frac{1}{-30 \text{ cm}} \quad \Rightarrow \quad f = -30 \text{ cm}$$

The focal length is half the radius of curvature, so

$$f = \frac{R}{2} \quad \Rightarrow \quad R = 2f = 2(-30 \text{ cm}) = -60 \text{ cm} \quad \Rightarrow \quad |R| = 60 \text{ cm}$$

9. **C** Since the image is projected onto a screen, it must be real, and therefore inverted. The magnification must be negative, so

$$M = -\frac{1}{4} \quad \Rightarrow \quad -\frac{s_i}{s_o} = -\frac{1}{4} \quad \Rightarrow \quad s_o = 4s_i$$

Because $s_i = 60$ cm, the object distance, s_o, must be 240 cm. Therefore,

$$\frac{1}{s_o} + \frac{1}{s_i} = \frac{1}{f} \quad \Rightarrow \quad \frac{1}{240 \text{ cm}} + \frac{1}{60 \text{ cm}} = \frac{1}{f} \quad \Rightarrow \quad \frac{1}{f} = \frac{1}{48 \text{ cm}} \quad \Rightarrow \quad f = 48 \text{ cm}$$

10. **B** A bi-concave lens is a diverging lens. Diverging lenses (like convex mirrors) have negative focal lengths and therefore cannot form real images. (Note that (D) is false; diverging lenses and convex mirrors always form diminished, virtual images, as you can verify using the mirror and magnification equations.)

Section II: Free Response

1. (a) Interference effects can be observed only if the light is coherent. Using two independent light sources at the slits in Barrier 2 would generate incoherent light waves. The set-up shown guarantees that the light reaching the two slits will be coherent.

   (b) Maxima are located at positions given by the equation $x_m = m\lambda L/d$, where m is an integer ($y_0 = 0$ is the central maximum). The first-order maximum for red light occurs at

$$x_{1, \text{ red}} = \frac{1 \cdot \lambda_{\text{red}} L}{d} = \frac{1 \cdot (750 \times 10^{-9} \text{ m})(4.0 \text{ m})}{0.50 \times 10^{-3} \text{ m}} = 0.006 \text{ m} = 6.0 \text{ mm}$$

and the first-order maximum for violet light occurs at

$$x_{1,\,violet} = \frac{1 \cdot \lambda_{violet} L}{d} = \frac{1 \cdot (400 \times 10^{-9}\ \text{m})(4.0\ \text{m})}{0.50 \times 10^{-3}\ \text{m}} = 0.0032\ \text{m} = 3.2\ \text{mm}$$

Therefore, the vertical separation of these maxima on the screen is $\Delta x = 6.0\ \text{mm} - 3.2\ \text{mm} = 2.8\ \text{mm}$.

(c) Maxima are located at positions given by the equation $x_m = m\lambda L/d$, where m is an integer. We therefore want to solve the equation

$$x_{m,\,violet} = x_{n,\,orange\text{-}yellow}$$
$$\frac{m\lambda_{violet} L}{d} = \frac{n\lambda_{orange\text{-}yellow} L}{d}$$
$$m\lambda_{violet} = n\lambda_{orange\text{-}yellow}$$
$$m(400\ \text{nm}) = n(600\ \text{nm})$$

The smallest integers that satisfy this equation are $m = 3$ and $n = 2$. That is, the third-order maximum for violet light coincides with the second-order maximum for orange-yellow light. The position on the screen (relative to the central maximum at $x = 0$) of these maxima is

$$x_{3,\,violet} = \frac{m\lambda_{violet} L}{d} = \frac{3(400 \times 10^{-9}\ \text{m})(4.0\ \text{m})}{0.50 \times 10^{-3}\ \text{m}} = 0.0096\ \text{m} = 9.6\ \text{mm}$$

(d) Within the glass, the wavelength is reduced by a factor of $n = 1.5$ from the wavelength in air. Therefore, the difference in path lengths, $d \sin \theta$, must be equal to $m(\lambda/n)$ in order for constructive interference to occur. This implies that the maxima are located at positions given by $x_m = m\lambda L/(nd)$. The distance between adjacent bright fringes is therefore

$$x_{m+1} - x_m = \frac{(m+1)\lambda L}{nd} - \frac{m\lambda L}{nd} = \frac{\lambda L}{nd} = \frac{(500 \times 10^{-9}\ \text{m})(4.0\ \text{m})}{1.5(0.50 \times 10^{-3}\ \text{m})} = 0.0027\ \text{m} = 2.7\ \text{mm}$$

2. (a) Beam 1 undergoes a 180° phase change when it reflects off the soap film. The refracted portion of the incident beam does not undergo a phase change upon reflection at the film/air boundary, because it strikes the boundary to a medium whose index is lower; therefore, Beam 2 does not suffer a 180° phase change.

(b) We're told that the criterion for destructive interference in this case is $\Delta \ell = 2T = m(\lambda/n)$, where n is the refractive index of the soap film and m is a whole number. Therefore, the criterion for *constructive* interference must be

$$\Delta \ell = 2T = (m + \tfrac{1}{2})\frac{\lambda}{n}$$

Alternatively, $\Delta \ell = 2T = \dfrac{k}{2} \cdot \dfrac{\lambda}{n}$, where k is an *odd* whole number.

3. (a) Since the image is formed behind the mirror, it is virtual.

(b) Virtual images are upright.

(c) Writing $s_i = -72$ cm (negative because the image is virtual), the magnification is

$$M = -\frac{s_i}{s_o} = -\frac{-72 \text{ cm}}{40 \text{ cm}} = 1.8$$

so the height of the image is $h_i = |m| \cdot h_o = 1.8(5 \text{ cm}) = 9$ cm.

(d) The mirror equation can be used to solve for the mirror's focal length:

$$\frac{1}{s_o} + \frac{1}{s_i} = \frac{1}{f} \Rightarrow \frac{1}{40 \text{ cm}} + \frac{1}{-72 \text{ cm}} = \frac{1}{f} \Rightarrow \frac{1}{90 \text{ cm}} = \frac{1}{f} \Rightarrow f = 90 \text{ cm}$$

Therefore, the mirror's radius of curvature is $R = 2f = 2(90 \text{ cm}) = 180$ cm.

(e)

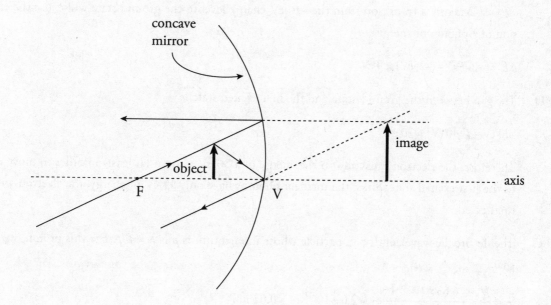

CHAPTER 11 REVIEW QUESTIONS

Section I: Multiple Choice

1. **B** Combining the equation $E = hf$ with $f = c/\lambda$ gives us

$$E = \frac{hc}{\lambda} = \frac{(4.14 \times 10^{-15}\ \text{eV} \cdot \text{s})(3.00 \times 10^8\ \text{m/s})}{2.07 \times 10^{-9}\ \text{m}} = 600\ \text{eV}$$

2. **D** The energy of the incident photons is

$E = hf = (4.14 \times 10^{-15}\ \text{eV} \cdot \text{s})(7.2 \times 10^{15}\ \text{Hz}) = 30\ \text{eV}$

Since $E > \phi$, photoelectrons will be produced, with maximum kinetic energy

$K_{max} = E - \phi = 30\ \text{eV} - 6\ \text{eV} = 24\ \text{eV}$

3. **A** If the atom's ionization energy is 25 eV, then the electron's ground-state energy must be −25 eV. Making a transition from the −16 eV energy level to the ground state will cause the emission of a photon of energy

$\Delta E = (-16\ \text{eV}) - (-25\text{eV}) = 9\ \text{eV}$

4. **D** The gap between the ground-state and the first excited state is

$-10\ \text{eV} - (-40\ \text{eV}) = 30\ \text{eV}$

Therefore, the electron must absorb the energy of a 30 eV photon (at least) in order to move even to the first excited state. Since the incident photons have only 15 eV of energy, the electron will be unaffected.

5. **C** The de Broglie wavelength of a particle whose momentum is p is $\lambda = h/p$. For this proton, we find that

$$\lambda = \frac{h}{p} = \frac{6.63 \times 10^{-34}\ \text{J} \cdot \text{s}}{3.3 \times 10^{-23}\ \text{kg} \cdot \text{m/s}} = 2.0 \times 10^{-11}\ \text{m} = 0.02\ \text{nm}$$

6. **D** If the atom begins in the $n = 3$ state, it could lose energy by making any of the following transitions: $3 \rightarrow 2$, $3 \rightarrow 1$, or $3 \rightarrow 2 \rightarrow 1$. The $3 \rightarrow 2$ transition would result in the emission of $-5\ \text{eV} - (-8\ \text{eV}) = 3\ \text{eV}$; the $3 \rightarrow 1$ transition would emit a $-5\ \text{eV} - (-12\ \text{eV}) = 7\ \text{eV}$ photon; and the $2 \rightarrow 1$ transition would result in the emission of $-8\ \text{eV} - (-12\ \text{eV}) = 4\ \text{eV}$. Therefore, if the atom is initially in the $n = 3$ state, it could emit photons of energy 3 eV, 4 eV, or 7 eV.

7. **C**   The energy of the nth level is given by the equation $E_n = E_1/n^2$, where E_1 is the ground-state energy. Therefore, the energy of level $n = 2$ (the next level above the ground state) is $E_2 = (-54.4\text{ eV})/4 = -13.6$ eV. Therefore, the difference in energy between the $n = 1$ and $n = 2$ levels is $\Delta E = -13.6\text{ eV} - (-54.4\text{ eV}) = 40.8$ eV.

8. **C**   The energy of a photon is given by the equation $E = hf$, or equivalently by $E = hc/\lambda$. Therefore, E is inversely proportional to λ. If λ decreases by a factor of 2, then E will increase by a factor of 2.

9. **A**   The equation that relates the mass difference m and the disintegration energy Q is Einstein's mass–energy equivalence formula, $Q = mc^2$. Because c^2 is a constant, we see that Q is proportional to m. Therefore, if m decreases by a factor of 4, then so will Q.

10. **B**   In order to balance the mass number (the superscripts), we must have $2 + 63 = 64 + A$, so $A = 1$. In order to balance the charge (the subscripts), we need $1 + 29 = 30 + Z$, so $Z = 0$. A particle with a mass number of 1 and no charge is a neutron, $_0^1\text{n}$.

11. **D**   In order to balance the mass number (the superscripts), we must have $196 + 1 = 197 + A$, so $A = 0$. In order to balance the charge (the subscripts), we need $78 + 0 = 78 + Z$, so $Z = 0$. The only particle listed that has zero mass number and zero charge is a gamma-ray photon, $_0^0\gamma$.

Section II: Free Response

1. (a)   If ionizing lithium twice leaves the atom with just one electron, then it must have originally had three electrons. Since neutral atoms have the same number of protons as electrons, Z must be 3 for lithium.

    (b)   First, use the formula given for E_n to determine the values of the first few electron energy levels:

    $$E_1 = \frac{3^2}{1^2}(-13.6\text{ eV}) = -122.4\text{ eV}$$

    $$E_2 = \frac{3^2}{2^2}(-13.6\text{ eV}) = -30.6\text{ eV}$$

    $$E_3 = \frac{3^2}{3^2}(-13.6\text{ eV}) = -13.6\text{ eV}$$

    $$E_4 = \frac{3^2}{4^2}(-13.6\text{ eV}) = -7.65\text{ eV}$$

    Differences between electron energy levels equal the energies of the emitted photons. Since $E = hf = hc/\lambda$, we have $\lambda = hc/E$. Since the E's are in electronvolts and the wavelengths in nanometers, it is particularly helpful to note that

    $$hc = (4.14 \times 10^{-15}\text{ eV} \cdot \text{s})(3.00 \times 10^8 \times 10^9\text{ nm/s}) = 1{,}240\text{ eV} \cdot \text{nm}$$

The goal here is to match the energy-level differences with the given wavelengths, using the equation $\lambda = (1{,}240 \text{ eV·nm})/E$.

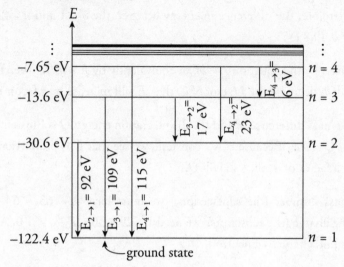

Using this diagram, the four wavelengths given correspond to the following photon energies and

$\lambda = 11.4 \text{ nm} \quad \leftrightarrow \quad E = \dfrac{hc}{\lambda} = \dfrac{1{,}240 \text{ eV} \cdot \text{nm}}{11.4 \text{ nm}} = 109 \text{ eV} \quad \leftrightarrow \quad (3 \rightarrow 1 \text{ transition})$

$\lambda = 13.5 \text{ nm} \quad \leftrightarrow \quad E = \dfrac{hc}{\lambda} = \dfrac{1{,}240 \text{ eV} \cdot \text{nm}}{13.5 \text{ nm}} = 92 \text{ eV} \quad \leftrightarrow \quad (2 \rightarrow 1 \text{ transition})$

$\lambda = 54.0 \text{ nm} \quad \leftrightarrow \quad E = \dfrac{hc}{\lambda} = \dfrac{1{,}240 \text{ eV} \cdot \text{nm}}{54.0 \text{ nm}} = 23 \text{ eV} \quad \leftrightarrow \quad (4 \rightarrow 2 \text{ transition})$

$\lambda = 72.9 \text{ nm} \quad \leftrightarrow \quad E = \dfrac{hc}{\lambda} = \dfrac{1{,}240 \text{ eV} \cdot \text{nm}}{72.9 \text{ nm}} = 17 \text{ eV} \quad \leftrightarrow \quad (3 \rightarrow 2 \text{ transition})$

(c)　No. In order for a spectral line to have a wavelength between 54.0 nm and 72.9 nm, there must be an energy-level transition with an energy between 17 eV and 23 eV. As the diagram above shows, no such intermediate transition is possible.

(d)　The 11.4 nm line corresponds to a photon energy of 109 eV. Therefore, the next shortest wavelength must correspond to the next highest possible photon energy. According to the diagram, the next highest available transition energy after 109 eV is 115 eV. This wavelength of the photon emitted in this $(4 \rightarrow 1)$ transition is

$$\lambda = \frac{hc}{E} = \frac{1{,}240 \text{ eV} \cdot \text{nm}}{115 \text{ eV}} = 10.8 \text{ nm}$$

Part VI
Practice Test 2

- Practice Test 2
- Practice Test 2: Answers and Explanations

Practice Test 2

AP® Physics 2 Exam

SECTION I: Multiple-Choice Questions

DO NOT OPEN THIS BOOKLET UNTIL YOU ARE TOLD TO DO SO.

At a Glance
Total Time
90 minutes
Number of Questions
50
Percent of Total Grade
50%
Writing Instrument
Pen required

Instructions

Section I of this examination contains 50 multiple-choice questions. Fill in only the ovals for numbers 1 through 50 on your answer sheet.

CALCULATORS MAY BE USED IN BOTH SECTIONS OF THE EXAMINATION.

Indicate all of your answers to the multiple-choice questions on the answer sheet. No credit will be given for anything written in this exam booklet, but you may use the booklet for notes or scratch work. After you have decided which of the suggested answers is best, completely fill in the corresponding oval on the answer sheet. Give only one answer to each question. If you change an answer, be sure that the previous mark is erased completely. Here is a sample question and answer.

Sample Question   Sample Answer

Chicago is a
(A) state
(B) city
(C) country
(D) continent

Use your time effectively, working as quickly as you can without losing accuracy. Do not spend too much time on any one question. Go on to other questions and come back to the ones you have not answered if you have time. It is not expected that everyone will know the answers to all the multiple-choice questions.

About Guessing

Many candidates wonder whether or not to guess the answers to questions about which they are not certain. Multiple choice scores are based on the number of questions answered correctly. Points are not deducted for incorrect answers, and no points are awarded for unanswered questions. Because points are not deducted for incorrect answers, you are encouraged to answer all multiple-choice questions. On any questions you do not know the answer to, you should eliminate as many choices as you can, and then select the best answer among the remaining choices.

GO ON TO THE NEXT PAGE.

ADVANCED PLACEMENT PHYSICS 2 TABLE OF INFORMATION

CONSTANTS AND CONVERSION FACTORS

Proton mass, $m_p = 1.67 \times 10^{-27}$ kg	Electron charge magnitude, $e = 1.60 \times 10^{-19}$ C
Neutron mass, $m_n = 1.67 \times 10^{-27}$ kg	1 electron volt, 1 eV $= 1.60 \times 10^{-19}$ J
Electron mass, $m_e = 9.11 \times 10^{-31}$ kg	Speed of light, $c = 3.00 \times 10^8$ m/s
Avogadro's number, $N_0 = 6.02 \times 10^{23}$ mol^{-1}	Universal gravitational constant, $G = 6.67 \times 10^{-11}$ m^3/kg·s^2
Universal gas constant, $R = 8.31$ J/(mol·K)	Acceleration due to gravity at Earth's surface, $g = 9.8$ m/s^2
Boltzmann's constant, $k_B = 1.38 \times 10^{-23}$ J/K	

1 unified atomic mass unit,	1 u $= 1.66 \times 10^{-27}$ kg $= 931$ MeV/c^2
Planck's constant,	$h = 6.63 \times 10^{-34}$ J·s $= 4.14 \times 10^{-15}$ eV·s
	$hc = 1.99 \times 10^{-25}$ J·m $= 1.24 \times 10^3$ eV·nm
Vacuum permittivity,	$\varepsilon_0 = 8.85 \times 10^{-12}$ C^2/N·m^2
Coulomb's law constant,	$k = 1/4\pi\varepsilon_0 = 9.0 \times 10^9$ N·m^2/C^2
Vacuum permeability,	$\mu_0 = 4\pi \times 10^{-7}$ (T·m)/A
Magnetic constant,	$k' = \mu_0/4\pi = 1 \times 10^{-7}$ (T·m)/A
1 atmosphere pressure,	1 atm $= 1.0 \times 10^5$ N/m$^2 = 1.0 \times 10^5$ Pa

UNIT SYMBOLS	meter,	m	mole,	mol	watt,	W	farad,	F
	kilogram,	kg	hertz,	Hz	coulomb,	C	tesla,	T
	second,	s	newton,	N	volt,	V	degree Celsius,	°C
	ampere,	A	pascal,	Pa	ohm,	Ω	electron volt,	eV
	kelvin,	K	joule,	J	henry,	H		

PREFIXES

Factor	Prefix	Symbol
10^{12}	tera	T
10^9	giga	G
10^6	mega	M
10^3	kilo	k
10^{-2}	centi	c
10^{-3}	milli	m
10^{-6}	micro	μ
10^{-9}	nano	n
10^{-12}	pico	p

VALUES OF TRIGONOMETRIC FUNCTIONS FOR COMMON ANGLES

θ	0°	30°	37°	45°	53°	60°	90°
$\sin\theta$	0	1/2	3/5	$\sqrt{2}/2$	4/5	$\sqrt{3}/2$	1
$\cos\theta$	1	$\sqrt{3}/2$	4/5	$\sqrt{2}/2$	3/5	1/2	0
$\tan\theta$	0	$\sqrt{3}/3$	3/4	1	4/3	$\sqrt{3}$	∞

The following conventions are used in this exam.
I. The frame of reference of any problem is assumed to be inertial unless otherwise stated.
II. In all situations, positive work is defined as work done on a system.
III. The direction of current is conventional current: the direction in which positive charge would drift.
IV. Assume all batteries and meters are ideal unless otherwise stated.
V. Assume edge effects for the electric field of a parallel plate capacitor unless otherwise stated.
VI. For any isolated electrically charged object, the electric potential is defined as zero at infinite distance from the charged object.

ADVANCED PLACEMENT PHYSICS 2 EQUATIONS

MECHANICS

$$v_x = v_{x0} + a_x t$$

$$x = x_0 + v_{x0}t + \frac{1}{2}a_x t^2$$

$$v_x^2 = v_{x0}^2 + 2a_x(x - x_0)$$

$$\vec{a} = \frac{\sum \vec{F}}{m} = \frac{\vec{F}_{net}}{m}$$

$$\left|\vec{F}_f\right| \leq \mu \left|\vec{F}_n\right|$$

$$a_c = \frac{v^2}{r}$$

$$\vec{p} = m\vec{v}$$

$$\Delta \vec{p} = \vec{F}\Delta t$$

$$K = \frac{1}{2}mv^2$$

$$\Delta E = W = F_{\parallel}d = Fd\cos\theta$$

$$P = \frac{\Delta E}{\Delta t}$$

$$\theta = \theta_0 + \omega_0 t + \frac{1}{2}\alpha t^2$$

$$\omega = \omega_0 + \alpha t$$

$$x = A\cos(\omega t) = A\cos(2\pi f t)$$

$$x_{cm} = \frac{\sum m_i x_i}{\sum m_i}$$

$$\vec{\alpha} = \frac{\sum \vec{\tau}}{I} = \frac{\vec{\tau}_{net}}{I}$$

$$\tau = r_{\perp}F = rF\sin\theta$$

$$L = I\omega$$

$$\Delta L = \tau \Delta t$$

$$K = \frac{1}{2}I\omega^2$$

$$\left|\vec{F}_s\right| = k\left|\vec{x}\right|$$

a	= acceleration
A	= amplitude
d	= distance
E	= energy
F	= force
f	= frequency
I	= rotational inertia
K	= kinetic energy
k	= spring constant
L	= angular momentum
ℓ	= length
m	= mass
P	= power
p	= momentum
r	= radius or separation
T	= period
t	= time
U	= potential energy
v	= speed
W	= work done on a system
x	= position
y	= height
α	= angular acceleration
μ	= coefficient of friction
θ	= angle
τ	= torque
ω	= angular speed

$$U_s = \frac{1}{2}kx^2$$

$$\Delta U_g = mg\Delta y$$

$$T = \frac{2\pi}{\omega} = \frac{1}{f}$$

$$T_s = 2\pi\sqrt{\frac{m}{k}}$$

$$T_p = 2\pi\sqrt{\frac{\ell}{g}}$$

$$\left|\vec{F}_g\right| = G\frac{m_1 m_2}{r^2}$$

$$\vec{g} = \frac{\vec{F}_g}{m}$$

$$U_G = -\frac{Gm_1 m_2}{r}$$

ELECTRICITY AND MAGNETISM

$$\left|\vec{F}_E\right| = \frac{1}{4\pi\varepsilon_0}\frac{\left|q_1 q_2\right|}{r^2}$$

$$\vec{E} = \frac{\vec{F}_E}{q}$$

$$\left|\vec{E}\right| = \frac{1}{4\pi\varepsilon_0}\frac{\left|q\right|}{r^2}$$

$$\Delta U_E = q\Delta V$$

$$V = \frac{1}{4\pi\varepsilon_0}\frac{q}{r}$$

$$\left|\vec{E}\right| = \left|\frac{\Delta V}{\Delta r}\right|$$

$$\Delta V = \frac{Q}{C}$$

$$C = \kappa\varepsilon_0\frac{A}{d}$$

$$E = \frac{Q}{\varepsilon_0 A}$$

$$U_C = \frac{1}{2}Q\Delta V = \frac{1}{2}C(\Delta V)^2$$

$$I = \frac{\Delta Q}{\Delta t}$$

$$R = \frac{\rho\ell}{A}$$

$$P = I\Delta V$$

$$I = \frac{\Delta V}{R}$$

$$R_s = \sum_i R_i$$

$$\frac{1}{R_p} = \sum_i \frac{1}{R_i}$$

$$C_p = \sum_i C_i$$

$$\frac{1}{C_s} = \sum_i \frac{1}{C_i}$$

$$B = \frac{\mu_0}{2\pi}\frac{I}{r}$$

A	= area
B	= magnetic field
C	= capacitance
d	= distance
E	= electric field
$\mathcal{E}$	= emf
F	= force
I	= current
ℓ	= length
P	= power
Q	= charge
q	= point charge
R	= resistance
r	= separation
t	= time
U	= potential (stored) energy
V	= electric potential
v	= speed
κ	= dielectric constant
ρ	= resistivity
θ	= angle
Φ	= flux

$$\vec{F}_M = q\vec{v} \times \vec{B}$$

$$\left|\vec{F}_M\right| = \left|q\vec{v}\right|\left|\sin\theta\right|\left|\vec{B}\right|$$

$$\vec{F}_M = I\vec{\ell} \times \vec{B}$$

$$\left|\vec{F}_M\right| = \left|I\vec{\ell}\right|\left|\sin\theta\right|\left|\vec{B}\right|$$

$$\Phi_B = \vec{B}\cdot\vec{A}$$

$$\Phi_B = \left|\vec{B}\right|\cos\theta\left|\vec{A}\right|$$

$$\mathcal{E} = -\frac{\Delta\Phi_B}{\Delta t}$$

$$\mathcal{E} = B\ell v$$

ADVANCED PLACEMENT PHYSICS 2 EQUATIONS

FLUID MECHANICS AND THERMAL PHYSICS

$$\rho = \frac{m}{V}$$

$$P = \frac{F}{A}$$

$$P = P_0 + \rho g h$$

$$F_b = \rho V g$$

$$A_1 v_1 = A_2 v_2$$

$$P_1 + \rho g y_1 + \frac{1}{2}\rho v_1^2$$
$$= P_2 + \rho g y_2 + \frac{1}{2}\rho v_2^2$$

$$\frac{Q}{\Delta t} = \frac{kA\,\Delta T}{L}$$

$$PV = nRT = Nk_B T$$

$$K = \frac{3}{2}k_B T$$

$$W = -P\,\Delta V$$

$$\Delta U = Q + W$$

A = area
F = force
h = depth
k = thermal conductivity
K = kinetic energy
L = thickness
m = mass
n = number of moles
N = number of molecules
P = pressure
Q = energy transferred to a system by heating
T = temperature
t = time
U = internal energy
V = volume
v = speed
W = work done on a system
y = height
ρ = density

WAVES AND OPTICS

$$\lambda = \frac{v}{f}$$

$$n = \frac{c}{v}$$

$$n_1 \sin\theta_1 = n_2 \sin\theta_2$$

$$\frac{1}{s_i} + \frac{1}{s_o} = \frac{1}{f}$$

$$|M| = \left|\frac{h_i}{h_o}\right| = \left|\frac{s_i}{s_o}\right|$$

$$\Delta L = m\lambda$$

$$d\sin\theta = m\lambda$$

d = separation
f = frequency or focal length
h = height
L = distance
M = magnification
m = an integer
n = index of refraction
s = distance
v = speed
λ = wavelength
θ = angle

GEOMETRY AND TRIGONOMETRY

Rectangle
$$A = bh$$

Triangle
$$A = \frac{1}{2}bh$$

Circle
$$A = \pi r^2$$
$$C = 2\pi r$$

Rectangular solid
$$V = \ell wh$$

Cylinder
$$V = \pi r^2 \ell$$
$$S = 2\pi r\ell + 2\pi r^2$$

Sphere
$$V = \frac{4}{3}\pi r^3$$
$$S = 4\pi r^2$$

A = area
C = circumference
V = volume
S = surface area
b = base
h = height
ℓ = length
w = width
r = radius

Right triangle
$$c^2 = a^2 + b^2$$

$$\sin\theta = \frac{a}{c}$$

$$\cos\theta = \frac{b}{c}$$

$$\tan\theta = \frac{a}{b}$$

MODERN PHYSICS

$$E = hf$$

$$K_{max} = hf - \phi$$

$$\lambda = \frac{h}{p}$$

$$E = mc^2$$

E = energy
f = frequency
K = kinetic energy
m = mass
p = momentum
λ = wavelength
ϕ = work function

AP PHYSICS 2

SECTION I

Time—90 minutes

50 Questions

Note: To simplify calculations, you may use $g = 10$ m/s^2 in all problems.

Directions: Each of the questions or incomplete statements below is followed by four suggested answers or completions. Select the one that is best in each case and mark it on your sheet.

1. A charged particle moves through a magnetic field and experiences a force F. New particles are sent into the same magnetic field. If the new particles have twice the charge, twice the mass, and twice the velocity, the new force would be

   (A) $4F$
   (B) $2F$
   (C) F
   (D) $1/2 F$

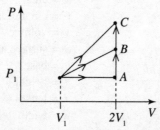

2. A cylinder with a piston on top contains an ideal gas at an initial pressure P_1, volume V_1, and Temperature T_1. The gas is allowed to expand to twice its original volume in three different ways as shown. The mechanical work done by the gas is

   (A) greatest for path A
   (B) greatest for path B
   (C) greatest for path C
   (D) equal for all paths

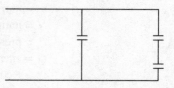

3. In the figure above, each capacitor has a capacitance of C. What is the overall capacitance of the above circuit?

   (A) 3 C
   (B) 3/2 C
   (C) 2/3 C
   (D) 1/3 C

4. You have a wire of length L, radius r, and resistance R. You need to obtain half that resistance using the same material and changing only one factor. You could

   (A) use half the length
   (B) use twice the length
   (C) use half the radius
   (D) use twice the radius

5. An electric motor has a label on it that reads: Input: 120V AC, 1.0 Amps, 60 Hz – Efficiency – 75%. At what constant speed can the motor lift up a 6 kg mass?

   (A) 0.5 m/s
   (B) 1.0 m/s
   (C) 1.5 m/s
   (D) 2.0 m/s

GO ON TO THE NEXT PAGE.

Questions 6-7 refer to the following figure:

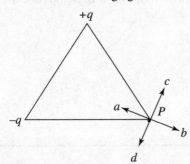

6. Two charges are placed as shown at the vertices of an equilateral triangle. What is the direction of the electric field at point *P* ?

(A) *a*

(B) *b*

(C) *c*

(D) *d*

7. A hollow conducting sphere is placed around point **P** so that the center is at **P**. What happens to the electric field strength at point **P** ?

(A) It doubles.

(B) It halves.

(C) It remains unchanged.

(D) It becomes zero.

8. A machine shoots a proton, a neutron, or an electron into a magnetic field at various locations. The paths of two particles are shown above. Assume they are far enough apart so that they do not intersect and the magnetic field is going out of the page. What can you say about the paths that represent each particle?

(A) *a* is the proton and *b* is the electron.

(B) *b* is the proton and *a* is the electron.

(C) Either may be a neutron.

(D) You cannot make any conclusions without knowing the velocities.

9. Heat is added to a cylindrical aluminum rod of radius *r* and length ℓ. The temperature difference between the two ends of the rod varies from 10°C to 20°C. What geometric factors will influence the rate heat is transferred along the rod?

(A) Only the length

(B) Only the area of the rod

(C) Both the length and area

(D) Neither the length nor area

10. A rigid, solid container of constant volume holds an ideal gas of volume V_1 and temperature T_1 and pressure P_1. The temperature is increased in an isochoric process. Which of the following is NOT true?

(A) The average speed of the molecules increases.

(B) The pressure increases.

(C) The kinetic energy of the system increases.

(D) The volume increases.

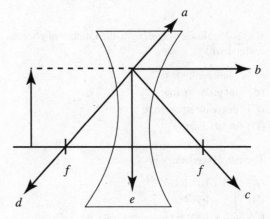

11. In the figure above, a ray of light hits an object and travels parallel to the principal axis as shown by the dotted line. Which line shows the correct continuation of the ray after it hits the concave lens?

(A) *a*

(B) *b*

(C) *c*

(D) *d*

GO ON TO THE NEXT PAGE.

12. A gas undergoes an expansion-compression cycle. If, plotted on a *P–V* diagram, the cycle is counterclockwise and the work is 300 J in magnitude, what was the heat transfer during this cycle?

    (A) 300 J into the system
    (B) 300 J out of the system
    (C) 600 J into the system
    (D) 600 J out of the system

13. Ultraviolet light has a wavelength of about 6×10^{-8} m. What is the frequency of this light?

    (A) 5×10^{15} Hz
    (B) 0.5 Hz
    (C) 2 Hz
    (D) 20 Hz

Questions 14-15 refer to the following situation:

Photons are emitted only when incoming radiation of wavelengths less than 2×10^{-6} m illuminates a metallic surface.

14. If you double the wavelength, the number of photons emitted will

    (A) increase by two
    (B) increase by four
    (C) decrease to one half
    (D) drop to zero

15. The cutoff frequency is

    (A) 2.0×10^{-6} Hz
    (B) 1.5×10^{14} Hz
    (C) 6.7×10^{-15} Hz
    (D) 4.0×10^{12} Hz

16. Which of the following is an isotope of Carbon $^{12}_{6}C$ with 7 neutrons in it?

    (A) $^{12}_{7}C$
    (B) $^{13}_{6}C$
    (C) $^{6}_{12}C$
    (D) $^{12}_{13}C$

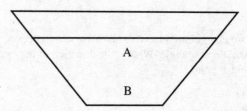

17. A cup of water is shaped as shown. You can say

    (A) the pressure is greater at point A due to the greater horizontal area
    (B) the pressure is greater at point A due to the smaller depth below the surface
    (C) the pressure is greater at point B due to the smaller horizontal area
    (D) the pressure is greater at point B due to the greater depth below the surface

GO ON TO THE NEXT PAGE.

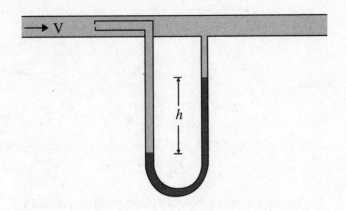

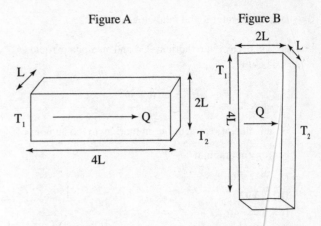

Figure A   Figure B

27. The instrument in an aircraft to measure airspeed is known as the pitot tube, shown in the figure above. The opening facing the incoming air (with the small aperature) is the part meant to capture the air at rest. The opening perpendicular to the flow of air (with the large aperature) is meant to capture air at speed. If $h = 1$ m and the fluid within the manometer is water, what is the airspeed? Take the density of air to be $\rho_{air} = 1.2$ kg/m^3.

(A) 27 m/s
(B) 68 m/s
(C) 95 m/s
(D) 128 m/s

28. When a light wave in air hits a thin film ($n_{film} > n_{air}$), part of the wave is transmitted into the film and part is reflected. At the bottom of the thin film part of the wave is transmitted into air and part is reflected back up. What can you say about the reflected wave's phase at the top layer and bottom layers?

(A) Only the top reflection is inverted.
(B) Only the bottom reflection is inverted.
(C) Neither reflection is inverted.
(D) Both reflections are inverted.

29. An object is placed 10 cm in front of a diverging mirror. What is the focal length of the mirror if the image appears 2 cm behind the mirror?

(A) –3/5 cm
(B) –5/3 cm
(C) –2/5 cm
(D) –5/2 cm

30. In the figure above, heat flows through a block of iron as shown in Figure A at a rate of H. The temperature on each side of the block are T_1 and T_2 respectively. If the block is rotated 90° as shown in Figure B the rate at which heat is transferred is now

(A) 1/4H
(B) H
(C) 2H
(D) 4H

31. Two charged, massive particles are isolated from all influence except those between the particle. They have charge and mass such that the net force between them is 0 N. Which of the following is NOT true?

(A) The particles must have the same sign of charge.
(B) If the distance between the particles changes, the force will no longer be 0 N.
(C) The particles must have the same mass.
(D) Such a situation is impossible as the net force between the two particles cannot be 0.

GO ON TO THE NEXT PAGE.

Questions 32-33 refer to the following situation:

Uranium-238 decays into thorium-234 and an alpha particle as illustrated below.

$$^{238}_{92}U \rightarrow ^{234}_{90}Th + ^4_2\alpha$$

32. What is the ratio of neutrons in thorium to the number of neutrons in uranium $\frac{n_{Th}}{n_U}$?

(A) 234/238
(B) 90/92
(C) 144/146
(D) 1/1

33. Which of the following statements is NOT true for the above reaction?

(A) Charge is conserved.
(B) The number of nucleons is conserved.
(C) Mass–energy is conserved.
(D) Energy is not conserved.

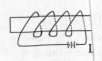 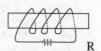

L   C   R

34. Three solenoids are arranged next to each other as shown above. What can we say about the solenoids interactions?

(A) L and attract; C and R repel.
(B) L and attract; C and R attract.
(C) L and repel; C and R repel.
(D) L and repel; C and R attract.

35. A particle mass M and charge q and velocity v is directed toward a uniform electric field of strength E and travels a distance d. How far does the particle travel if the original velocity is doubled and the mass is cut in half?

(A) $4d$
(B) $2d$
(C) d
(D) $1/2d$

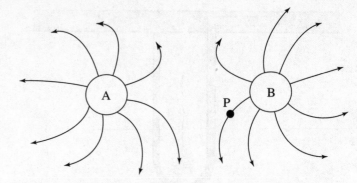

36. The diagram above shows the electric field lines around a region of space due to two charges A and B. Which of the following statements is true?

(A) The field is uniform.
(B) Charge A is positive and charge B is negative.
(C) A positive charge starting at rest from point P would move toward B.
(D) The equipotential lines are perpendicular to the field lines.

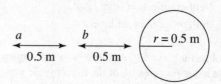

37. A particle travels through a uniform electric field $E = 5$ N/C as shown. What is the size and direction of the magnetic field needed to have a 0.2 C charge with mass 0.005 kg pass through undeflected at a velocity of 4 m/s?

(A) 20 T out of the page
(B) 20 T into the page
(C) 1.2 T out of the page
(D) 1.2 T into the page

<table>
<tr><td>a</td><td>b</td><td>r = 0.5 m</td></tr>
<tr><td>0.5 m</td><td>0.5 m</td><td></td></tr>
</table>

38. A hollow sphere carries a charge of 3 C. How does the magnitude of the field at point a compare to point b ?

(A) $E_a = (1/4) E_b$
(B) $E_a = (4/9) E_b$
(C) $E_a = (9/4) E_b$
(D) $E_a = 4 E_b$

GO ON TO THE NEXT PAGE.

39. A solid plastic cube with uniform density (side length = 0.5 m) of mass 100 kg is placed in a vat of fluid whose density is 1200 kg/m³. What fraction of the cube's volume floats above the surface of the fluid?

    (A) 1/9
    (B) 1/3
    (C) 2/3
    (D) 4/5

40. A point charge, $Q = +1$ mC, is fixed at the origin. How much work is required to move a charge, $q = +8$ μC, from the point (0, 4 meters) to the point (3 meters, 0)?

    (A) 3.5 J
    (B) 6.0 J
    (C) 22.5 J
    (D) 40 J

41. The following three steps process refers to a simple RC circuit with a battery and an initially open switch.

    Step 1: The switch is closed, allowing the capacitor to charge.
    Step 2: After the capacitor has charged, a slab of dielectric material is inserted between the plates of the capacitor and time passes.
    Step 3: The switch is opened, and the dielectric is removed.

    Which of the following describes the change in potential energy stored in the capacitor during each step?

    (A) $\Delta U_1 < 0; \Delta U_2 > 0; \Delta U_3 > 0$
    (B) $\Delta U_1 > 0; \Delta U_2 > 0; \Delta U_3 > 0$
    (C) $\Delta U_1 < 0; \Delta U_2 > 0; \Delta U_3 < 0$
    (D) $\Delta U_1 > 0; \Delta U_2 < 0; \Delta U_3 < 0$

42. An electron (mass = m, charge = $-e$) is rotated with speed v upwards, in the plane of the page, into a region containing a uniform magnetic field **B**, that is directed into the plane of the page. Describe the electron's subsequent circular motion.

    (A) Clockwise rotation; radius of path = $mv/(e\text{B})$
    (B) Counterclockwise rotation; radius of path = $mv/(e\text{B})$
    (C) Clockwise rotation; radius of path = $e\text{B}/(mv)$
    (D) Counterclockwise rotation; radius of path = $e\text{B}/(mv)$

Questions 43-44 refer to the following figure:

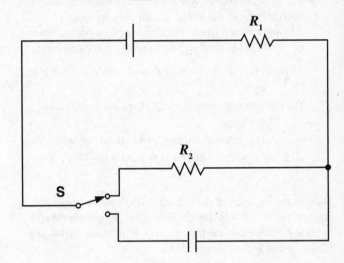

43. When switch is in the position shown in the figure, what is the current through the circuit? The voltage of the battery is 10 V, $R_1 = 2\ \Omega$ and $R_2 = 3\ \Omega$.

    (A) 2.0 A
    (B) 4.1 A
    (C) 6.5 A
    (D) 8.3 A

44. When the switch is moved to the opposite position, which of the following statements descirbed the current through the circuit. Consider the capacitor as starting with no charge.

    (A) The current starts at 0 and grows to a maximum of 2 A.
    (B) The current starts at 0 and grows to a maximum of 5 A.
    (C) The current starts at a maximum of 2 A and drops to 0.
    (D) The current starts at a maximum of 5 A and drops to 0.

GO ON TO THE NEXT PAGE.

45. Optometrists use a linear scale of lens power, measured in diopters, to prescribe corrective lenses. Sufferers of myopia, or nearsightedness, have difficulty resolving distant objects because the lenses in their eyes are too strong. If a myopic person has a prescription of –5 diopters, what image do her corrective lenses create for distant objects?

   (A) An upright, real image about 20 cm in front of her eyes

   (B) An upright, virtual image about 20 cm in front of her eyes

   (C) An inverted, real image about 20 cm behind her eyes

   (D) An inverted, virtual image 5 cm behind her eyes

Directions: For each of the questions 46-50 below, <u>two</u> of the suggest answers will be correct. Select the two answers that are best in each case, and then fill in both of the corresponding circles on the answer sheet.

46. There is a negatively charged hollow sphere. Which of the following statements can be said of the sphere? Select two answers.

   (A) The electric field on the inside is zero.

   (B) For some distance $r > r_{sphere}$ you can consider the charge to be concentrated at the center of the sphere.

   (C) For some distance $r > r_{sphere}$ you can consider the charge to be concentrated on the outside of the sphere.

   (D) The electric field points outward away from the center of the sphere.

47. Which of the following statements is true for a diverging lens? Select two answers.

   (A) It can only form virtual images.

   (B) It can only form real images.

   (C) The magnification is always less than 1.

   (D) It can form real images if the object is beyond the focal length and virtual images if the object is inside the focal length.

48. Which of the following statements is true for both sound and light waves? Select two answers.

   (A) They can both be polarized.

   (B) They both do not require mediums to travel.

   (C) They can both diffract.

   (D) They can both refract.

49. Which of the following statements is true for lenses? Select two answers.

   (A) The focal length of a lens depends upon the radius of curvature of the lens.

   (B) The focal length of the lens depends on the light intensity.

   (C) The focal length of the lens depends on the shape of the lens.

   (D) The focal length is directly proportional to the power of the lens.

50. N resistors ($N > 2$) are connected in parallel with a battery of voltage V_0. If one of the resistors is removed from the circuit, which of the following quantities will decrease? Select two answers.

   (A) The voltage across any of the remaining resistors

   (B) The current output by the battery

   (C) The total power dissipated in the circuit

   (D) The total resistance in the circuit

END OF SECTION I

AP PHYSICS 2

SECTION II

Time—90 minutes

4 Questions

Directions: Questions 1 and 2 are long free-response questions that require about 30 minutes to answer. Questions 3 and 4 are short answer questions that require about 15 minutes to answer. Show your work for each part in the space provided after that part.

1. In an experiment conducted, two tests are run. In both trials you may ignore the effect of gravity. The following is a diagram of the tests.

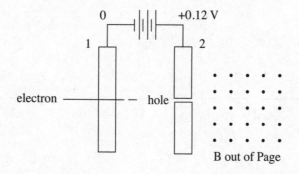

Test 1: There are two large parallel plates separated by a distance $d = 0.5$ m with a potential difference of 0.12 V across them. There is a uniform magnetic field **B** pointing perpendicularly out of the paper of strength 0.002 T starting to the right of plate 2. An electron is released from rest at plate 1 as shown above. It passes through a hole in plate 2 and enters the magnetic field and only experiences forces due to the magnetic field.

Test 2: The same set-up is run with the following two exceptions.

1. The battery is switched so that plate 1 becomes positive and plate 2 becomes negative.

2. A proton is used instead of an electron.

(a) Compare the magnitudes of force acting on each charge.

(b) Compare the speed of the proton as it emerges from the hole to the speed of the electron as it emerges from the hole.

(c) The charges will curve around once they are in the magnetic field. Make a sketch of each path. How far away will the two charges be when they strike the plate?

(d) Compare the time it takes for the electron to return to the plate to the time it takes the proton to return to the plate.

GO ON TO THE NEXT PAGE.

2. Three charges are fixed at positions along the x-axis as positions $-d$, 0, and $+d$. The charges at $-d$ and $+d$ are both negative, and the charge at 0 is positive.

    (a) A positively charged object of mass m is placed on the x-axis between 0 and $+d$, close to the position $x = 0$. If the three charges described above do not move as a result of this new charged object, describe the motion of object after it is released as it moves in the region $0 < x < d$.

The charge at 0 has a magnitude of $2Q$, while the other two charges have a magnitude of Q.

    (b) On the axes below, sketch the electric field along the x-axis in the vicinity of the charges. An electric field to the right should be graphed as positive and a field pointing left should be graphed as negative.

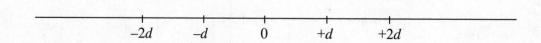

                        $-2d$ $-d$ 0 $+d$ $+2d$

    (c) Write a mathematical function, $E(x)$, that gives the electric field at any position along the x-axis for $0 < x < d$. Give your answer in terms of Q, d, and fundamental constants.

    (d) In order to originally assemble the three charges on the x-axis, some work had to be done. Consider arranging the charges along the x-axis in the following manner: first, bring the $+2Q$ charge to position $x = 0$, then bringing the $-Q$ charge to $x = +d$, and finally, bring in the last charge. Bringing the $+2Q$ charge to position 0 required no work. Bringing in the second charge required an amount of work W. Explain whether bringing in the third charge will require more work, less work, or an amount of work equal to W.

GO ON TO THE NEXT PAGE.

3. A student has a convex lens of unknown focal length. He lights a candle in a darkened room and uses the lens and moves a screen until he forms a sharp image and records the distance from the candle to the lens and the distance from the lens to the screen. Below is a sketch of his set-up and his data.

s_o (cm)	s_i (cm)
15	61.5
20	29.8
30	19.3
40	18.0
50	15.6
60	15.3
70	14.7
80	13.9
90	13.7
100	13.7
110	13.4
120	13.2

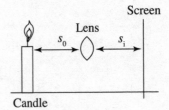

(a) What type of image is formed and what orientation is it?

(b) Where is the image formed if the candle is placed 6 cm from the lens?

(c) Use ray tracing to make a sketch when the object is 6 cm from the lens.

(d) What effect would it have if a lens with the same focal length was used but with only half the diameter?

GO ON TO THE NEXT PAGE.

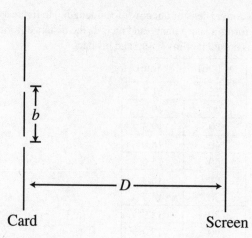

Note: Figure not drawn to scale.

4. In a double-slit interference experiment, a parallel beam of monochromatic light is needed to illuminate two narrow parallel slits of width w that are a distance h apart in an opaque card as shown in the figure above. A lens is inserted between the point light sources S and the slits in order to produce the parallel beams of light. The interference pattern is formed on a screen a distance D from the slits, where $D \gg h$.

    (a)   On the figure above, draw the lens at the appropriate place to produce the parallel beams of light, and label the location of the source relative to the lens with the appropriate optical parameters of the lens.

    b)   Draw two light rays from the source to the slits to show the production of parallel rays.

    (c)   In the interference patterns on the screen, the distance from the central bright fringe to the third bright fringe on one side is measured to be y_3. Derive an expression for the wavelength of the light in terms of the given quantities and fundamental constants.

    (d)   If the space between the slits and the screen was filled with a material having an index of refraction $n > 1$, would the distance between the bright fringes increase, decrease, or remain the same? Explain your reasoning.

STOP

END OF EXAM

Practice Test 2:
Answers and
Explanations

PRACTICE TEST 2 ANSWER KEY

1.	A	26.	D	
2.	C	27.	D	
3.	B	28.	A	
4.	A	29.	D	
5.	C	30.	D	
6.	D	31.	A	
7.	D	32.	C	
8.	A	33.	D	
9.	C	34.	D	
10.	D	35.	B	
11.	A	36.	D	
12.	B	37.	C	
13.	A	38.	B	
14.	D	39.	B	
15.	B	40.	B	
16.	B	41.	B	
17.	D	42.	A	
18.	B	43.	A	
19.	A	44.	D	
20.	A	45.	B	
21.	C	46.	A, B	
22.	C	47.	A, C	
23.	A	48.	C, D	
24.	D	49.	A, C	
25.	C	50.	B, C	

SECTION I

1. **A** The magnitude of the force is given by $F = qvB$. Doubling the charge doubles the force, doubling the velocity doubles the force, although doubling the mass will reduce the acceleration, it will not affect the force.

2. **C** The work done by a gas is determined by the area under a P vs. V graph. Because path C has the greatest area it does the greatest amount of work.

3. **B** First examine the two capacitors in series. The capacitors in series follow the rule $\frac{1}{C_S} = \sum_i \frac{1}{C_i}$. For two capacitors of equal value in parallel, the total capacitance is reduced by two. Once the series section is taken care of, the two branches in parallel can be determined by $C_P = \sum_i C_i$. This means the two branches can simply add. $1\,C + 1/2\,C = 3/2\,C$.

4. **A** The resistance is given by $R = \frac{\rho \ell}{A}$. Using the same material means ρ is constant. Cutting the length in half will reduce the resistance by a factor of 2. Cutting the radius in half or doubling the radius will change the resistance by a factor of 1/4 (1/4 or 4) because $A = \pi r^2$. The only correct answer is (A).

5. **C** $P = IV$ means the electrical power is $P = (1A)(120V) = 120$ W. If this motor is only 75% efficient we get $(0.75)(120$ W$) = 90$ W of power output. Also $P = Fv$. Lifting a 6 kg mass upward at a constant velocity requires a 60 N force (by $F_g = mg \rightarrow (6$ kg$)(10$ m/s$^2)$. Therefore $v = P/F$ or $v = 90$ W$/60$ N $= 1/5$ m/s.

6. **D** The two electric fields vectors are shown. The resultant can be determined as shown below.

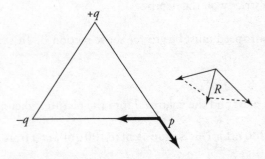

7. **D** Within a conducting sphere, the electric field is always zero. Therefore, in the region of P, as well as everywhere else within the conducting sphere, there is zero electric field.

8. **A** Path a follows the right-hand rule so must be positively charged. Path b is opposite the right hand rule so must be negatively charged. Neutrons would follow a straight line path, so it is impossible for either a or b to be a neutron.

9. **C** The rate of heat flow through a material is directly proportional to the area of the material and inversely proportional to the length of the material.

10. **D** Isochoric means constant volume. Also the average speed of the molecules is given by $v_{rms} = \sqrt{\dfrac{3RT}{M}}$ so increasing T increases the speed and therefore the number of collisions. From the ideal gas law $PV = nRT$ so keeping a constant volume means increasing T also increases P. $K_{ave} = \dfrac{3}{2}k_b T$ tells us the kinetic energy increases as the temperature does.

11. **A** The rules for ray-tracing diagrams state that a line parallel to the principle axis bends away from the focal point as shown.

12. **B** For a cycle, $\Delta U = 0$, so we know the magnitude of the work and heat transfer have to be the same. This eliminates (C) and (D). Since the cycle is counterclockwise on a P–V diagram, the work has to be positive, so the heat transfer is negative. A negative heat transfer is heat leaving the system. Choice (B) is correct.

13. **A** $v = f\lambda$ so $\dfrac{v}{\lambda} = f$. $f = \dfrac{3 \times 10^8 \text{ m/s}}{6 \times 10^{-8} \text{ m}} \rightarrow 0.5 \times 10^{16} \text{ Hz} = 5.0 \times 10^{15} \text{ Hz}$

14. **D** If you double the wavelength you reduce the frequency by a factor of two. As soon as you are below the threshold frequency no photons are emitted.

15. **B** Given $v = f\lambda$ we know $f = \dfrac{v}{\lambda} \rightarrow \dfrac{3 \times 10^8 \text{ m/s}}{2 \times 10^{-6} \text{ m}}$

16. **B** The lower number must remain 6 for carbon because the number of protons is equivalent to the atomic number. The upper left number is the sum of the protons and neutrons (6 + 7 = 13).

17. **D** The pressure in a non-moving fluid depends strictly on the depth.

18. **B** Because the fluid is higher in section B, the air speed must be greater above section B. This must be due to a narrower tube.

19. **A** To determine the density you need both the mass and the volume. From the picture, when the ball floats it displaces 100 mL of water (300 mL – 200 mL). This is equivalent to 100 g of water from $m = \rho V$ and the density of water is 1 g/mL. This weighs just about 1 N (from $F_g = mg \rightarrow F_g = 0.1 \text{ kg}(10 \text{ m/s}^2)$). When an object floats the buoyant force is equal to the weight of the object so the object weighs 1 N, which means the mass of the object is 100 g. When an object is submerged the object displaces an amount of water equal to the object's volume. Therefore, the volume of the object is 300 mL (500 mL – 200 mL). The density is now given by $\rho = \dfrac{m}{V} \rightarrow \dfrac{100 \text{ g}}{300 \text{ mL}} \rightarrow 0.33 \text{ g/mL}$.

20. **A** We know the work done by the electric field moving the charge q across the potential difference ΔV is going to be

$$W = -\Delta U_E$$

However, the work done to combat this electric force, in order to move the charge, is just the opposite sign of W, so

$$W = +q\Delta V = q(V_2 - V_1), \text{ which is (A).}$$

21. **C** The work done causes a change in kinetic energy. Because the particle starts from rest $W = \Delta K$. This becomes $q(V_2 - V_1) = \frac{1}{2}mv^2 \rightarrow \frac{2q(V_2 - V_1)}{m} = v^2 \rightarrow v = \sqrt{\frac{2q(V_2 - V_1)}{m}}$.

22. **C** Conventional current flows out of the positive side of the battery (the longer (left) side on the battery schematic). Therefore, the current in A is CCW. The right-side section of wire in A produces a magnetic field on the right of the loop that is into the page as given by the right-hand rule. Therefore, the flux in loop B is into the page. As loop B moves away from loop A, the flux into the page decreases. From Lenz's Law a current will be produced in loop B in order to oppose this decrease in flux. This would require a CW current as given by the right-hand rule.

23. **A** The voltage across the 6,000 Ω resistor is given by $V = IR$ or (0.001 A)(6,000 Ω) = 6 V. In a parallel circuit, the voltage across both branches is the same and so the total voltage drop across R1 and the 1,000 Ω resistor must also be 6 V. The voltage across R1 and the 1,000 Ω will add to 6 V because they are in series. Because there is 4 V across the 1,000 Ω resister, there must be 2 V.

24. **D** The current through the 1000 Ω resistor is given by $I = V/R$ or 4 V/1000 Ω = 0.004 A. The current of the two branches will sum up to the current flowing out of the battery in a parallel circuit. $I_B = 0.004$ A + 0.001 A = 0.005 A or 5 mA.

25. **C** The larger the angle when measured from the normal, the smaller the index of refraction. Because the angle in A > angle in C > angle in B, $n_A < n_C < n_B$.

26. **D** Water has a higher index of refraction than air, so if light refracts through the water the light will bend away from the normal—so C is possible. Another possibility is D, total internal reflection when the angle in water is greater than the critical angle.

27. **D** First, we will use Bernoulli's Equation to find the difference in pressure between the end of the tube capturing air at rest and the end capturing the air at speed. Let's call these ends A and B. Bernoulli's Equation tells us

$$P_A + \frac{1}{2}\rho_{air}v_A^2 + \rho_{air}gy_A = P_B + \frac{1}{2}\rho_{air}v_B^2 + \rho_{air}gy_B$$

Since end B is the end that captures the air at rest, $v_B = 0$. Looking at the figure, the heights of the two ends are basically the same, or $y_A \approx y_B$. So, we get

$$P_A - P_B = \frac{1}{2}\rho_{air}v_B^2$$

Now, in the manometer, we know that the pressure on the fluid near end A is just P_A, and the pressure on the fluid near end B is just P_B, so the difference in the two pressures must equal the weight of the fluid of height h, or

$$P_A - P_B = \rho_f g h$$

Putting these two things together, we get

$$\rho_f g h = \frac{1}{2} \rho_{air} v_B^2$$

$$\rightarrow v_B^2 = 2 \frac{\rho_f}{\rho_{air}} g h$$

$$\rightarrow v_B = \sqrt{2 \frac{\rho_f}{\rho_{air}} g h} = \sqrt{2 \frac{1,000 \ kg/m^3}{1.2 \ kg/m^3} (9.8 \ m/s^2)(1 \ m)} = 128 \ m/s$$

So, the airspeed is 128 m/s.

28. **A** Reflections are inverted when the new medium has a higher index of refraction. Reflections are not inverted when the new medium has a lower index of refraction. Because $n_{film} > n_{air}$ there is an inversion on the top, but not the bottom.

29. **D** The thin lens equation is given by $\frac{1}{s_i} + \frac{1}{s_o} = \frac{1}{f}$. Images formed behind a mirror have a negative in front of them: $\frac{1}{-2 \ cm} + \frac{1}{10 \ cm} = \frac{1}{f} \rightarrow \frac{-5}{10 \ cm} + \frac{1}{10 \ cm} = \frac{1}{f} \rightarrow \frac{-4}{10 \ cm} = \frac{1}{f}$ or $f = -5/2$ cm. This

answer makes sense because the focal length of a diverging mirror is negative.

30. **D** The rate at which heat is transferred is given by $H = \frac{kA\Delta T}{L}$. The material does not change, nor does

the change in temperature, so k and ΔT won't affect the answer. In figure a, the area is (L)(2L) or $2L^2$.

The length is 4L. This means the initial rate is $H = \frac{k2L^2\Delta T}{4L} \rightarrow H = \frac{kL\Delta T}{2}$. In figure b the area is

given by (L)(4L) or $4L^2$ and the length is 2L. The heat flow for b is $H = \frac{k4L^2\Delta T}{2L} \rightarrow H = 2kL\Delta T$.

Figure b has four times the rate as figure a.

31. **A** It is possible for the gravitation force to balance the electrical force, making (D) false. In order for this to occur, since gravity is always an attractive force, the electrical force must repulse, and therefore the charges must have the same sign, making (A) true. Both the gravitational and electrical force are inverse square laws, so if the distance between the particles changes, both forces change by the same factor, making (B) false. And (C) is just false, both particles will experience the same gravitational force regardless of how the mass is distributed amongst them; the gravitation force depends on the product of the two masses.

32. **C** The number of neutrons in an element can be found by taking the atomic mass minus the atomic number. For Thorium this becomes $234 - 90 = 144$, and for Uranium this becomes $238 - 92 = 146$. The ratio is therefore 144/146.

33. **D** All conservation rules are obeyed. Charge is conserved $(92 - 90 + 2)$. The number of nucleons is conserved (92 protons + 146 neutrons) = (90 protons + 144 neutrons) + (2 protons + 2 neutrons). Mass–energy is conserved $(238 = 234 + 4)$. Energy isn't conserved because it changes when mass is converted to energy.

34. **D** The right-hand rule lets us determine that the solenoids will behave like magnets as shown.

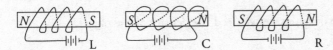

Because south poles of magnets repel south poles, and they attract north poles of magnets, the answer is (D).

35. **B** The work done moving a particle against an electric field is given by $W = qEd$. Work is also equal to the change in kinetic energy. Because the final velocity is zero, we can say $\frac{1}{2}mv^2 = qEd \rightarrow d = \frac{mv^2}{2qE}$. Cutting the mass in half cuts the distance in half but doubling the velocity makes the distance 4 times bigger. These factors combines to give the answer, (B).

36. **D** The field lines are neither evenly spaced nor parallel so the field is not uniform. The field lines point away from A and B and so both charges are positive and a positive charge would move in the direction of the field line away from B. Choice (D) is true.

37. **C** The force acting on a particle moving through an electric field is given by $F = qE$. The force acting on a charge moving through a magnetic field is equal to $F = qvB$. Setting these two forces equal to each other we get $qE = qvB$ or $E = vB$. Therefore, $B = \frac{E}{v} \rightarrow \frac{5 \text{ N/C}}{4 \text{ m/s}} \rightarrow 1.2$ T. To determine the direction, we know the force acting on the particle must be downward to counteract the upward push due to the electric field. Using the right-hand rule, we know the magnetic field must point out of the page.

38. **B** The electric field is given by $E = \dfrac{kq}{r^2}$ where r is measured from the center of the sphere to the point in question.

$$E_a = \frac{kq}{\left(\frac{3}{2}\right)^2} \rightarrow \frac{kq}{\frac{9}{4}} \rightarrow \frac{4kq}{9}$$

$$E_b = \frac{kq}{1^2} \rightarrow kq$$

Substituting E_b for the kq in the first equation becomes $E_a = \dfrac{4E_b}{9}$.

39. **B** We need to find the ratio of the densities, so we need the density of the object.

Volume $= V = (0.5 \text{ m})^3 = 0.125 \text{ m}^3$

$\rho_{object} = m/V = 100 \text{ kg}/0.125 \text{ m}^3 = 800 \text{ kg/m}^3$

$$\frac{V_{sub}}{V_{obj}} = \frac{\rho_{obj}}{\rho_{fluid}} = \frac{800 \text{ kg/m}^3}{1200 \text{ kg/m}^3} = \frac{2}{3}$$

2/3 is submerged, meaning 1/3 sits above the surface.

40. **B** Electricity is conservative. $\Delta U = qV$ and $\Phi = k\,(Q/r)$ so

$$V_{i \rightarrow f} = \Phi_i - \Phi_f = \left(9 \times 10^9 \text{ N} \frac{\text{m}^2}{\text{C}^2}\right)\left(\frac{1 \times 10^{-3} \text{ C}}{4 \text{ m}}\right) - \left(9 \times 10^9 \text{ N} \frac{\text{m}^2}{\text{C}^2}\right)\left(\frac{1 \times 10^{-3} \text{ C}}{3 \text{ m}}\right)$$

$$= -\frac{3}{4} \times 10^6 \text{ J/C}$$

Now for our potential energy,

$$\Delta U = qV = \left(-8 \times 10^{-6} \text{ C}\right)\left(-\frac{3}{4} \times 10^6 \frac{\text{J}}{\text{C}}\right) = 6.0 \text{ J}$$

41. **B** *Step 1:* Potential energy rises because the magnitude of charge on each plate increase, $Q_C = CV_C$ and $U_C = \dfrac{1}{2} CV^2$. $\Delta U_1 > 0$, eliminate (A) and (C).

Step 2: Inserting a dielectric increases potential energy. Inserting a dielectric always increases capacitance. $Q_C = CV_C$ and $U_C = \dfrac{1}{2} CV^2$, therefore, potential energy increases. $\Delta U_2 > 0$, so eliminate (D).

Step 3: Capacitor is still charged so there is still potential energy.

42.   **A**   By the right-hand rule the electron exhibits clockwise rotation. Eliminate (B) and (D). The centripetal force is provided by the magnetic force:

$$F_B \rightarrow F_C$$
$$qvB = \frac{mv^2}{r}$$
$$r = \frac{mv}{qB} = \frac{mv}{eB}$$

43.   **A**   When the switch is in the position given in the figure, you have a simple circuit with two resistors in series. The equivalent resistance of the circuit is therefore $R_{eq} = R_1 + R_2 = 5\ \Omega$, and the current in the circuit is

$$i = \frac{V}{R_{eq}} = \frac{10\ \text{V}}{5\ \Omega} = 2\ \text{A}$$

44.   **D**   When the switch is in the opposite position, then the circuit is an RC circuit. In this case, the capacitor is initially uncharged, so the circuit starts out as if there is no capacitor, with a maximum current that decreases. This eliminates (A) and (B). The maximum current is just the current that the circuit has when the capacitor isn't even there, or $i = 10\ \text{V}/2\ \Omega = 5$ A. This eliminates (C).

45.   **B**   Lens power is given by

$$P = 1/f \rightarrow f = 1/P = -1/5 = -0.2\ \text{m} = -20\ \text{cm}.$$

This eliminates (D). Because of the negative sign, the image sits in front of her eyes. Eliminate (C). The image sits in front of her eyes, meaning the image is virtual.

46.   **A, B**   In a hollow conducting sphere there is no electric field inside of it. When we move away from the sphere we can treat the conducting sphere as a point charge. Treating it as a point charge means we can consider the charge to be concentrated at the center of the sphere.

47.   **A, C**   Choices (A) and (C) are true for diverging lenses. Choice (D) is for a converging lens. Diverging lenses cannot produce real images.

48.   **C, D**   Because sound is a longitudinal wave, it cannot be polarized. Both longitudinal waves (sound) and transverse waves (light) can diffract (bending and spreading as it passes through an opening). Both can also refract (change speeds as the wave enters a new material). Electromagnetic waves (light) can travel through a vacuum.

49.   **A, C**   The intensity of the light does not affect local length. Power is given by $P = 1/f$. Power P has an inverse relationship with focal length f.

50. **B, C** The resistors are all in parallel, meaning their voltage is the same across each resistor and matches the voltage of the battery. Eliminate (A). The voltage provided by the battery does not increase or decrease; the battery still maintains the same voltage. Eliminate (D). The resistors are in parallel; eliminating one of the resistors will increase the equivalent resistance (there is now one less path for current to go through so equivalent resistance goes up). Because $I = V/R$ this increase in resistance would decrease the current. Power's equation is $P = IV$. If our current decreases, our power decreases as well.

SECTION II

1. (a) The electric field is given by both

$$E = \frac{F}{q} \text{ and } E = \frac{V}{d}$$

Combining these equations we get

$$\frac{F}{q} = \frac{V}{d}$$

$$F = \frac{qV}{d} = \frac{(1.6 \times 10^{-19} \text{ C}) \times (0.12 \text{ V})}{0.5 \text{ m}} = 3.8 \times 10^{-20} \text{ N}$$

Because the charges have the same magnitude and only the opposite direction, the voltage is the same except for the opposite direction, and the distance is the same, the charges will experience equal forces.

(b) Because we know the force and distance, we can use

$$W = \Delta K$$
$$Fd = \frac{1}{2}mv^2$$
$$\frac{2Fd}{m} = v^2$$

For the electron,

$$v_e = \sqrt{\frac{2Fd}{m_e}} = \sqrt{\frac{2(3.8 \times 10^{-20} \text{ N})(0.5 \text{ m})}{9.1 \times 10^{-31} \text{ kg}}} = 2.0 \times 10^5 \text{ m/s}$$

For the proton,

$$v_p = \sqrt{\frac{2Fd}{m_p}} = \sqrt{\frac{2(3.8 \times 10^{-20} \text{ N})(0.5 \text{ m})}{1.67 \times 10^{-27} \text{ kg}}} = 4.8 \times 10^3 \text{ m/s}$$

(c) The speed of the proton is only about 2% the speed of the electron. The charges will experience a magnetic force once they enter the magnetic field given by $F = qvB$. The force is perpendicular to the direction of motion so that the charges also obey the circular motion equation $F = mv^2/r$. A sketch would look like:

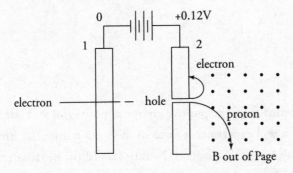

Setting these two forces equal to each other we have:

$$qvB = \frac{mv^2}{r}$$

$$qB = \frac{mv}{r}$$

$$r = \frac{mv}{qB}$$

A proton will have a radius given by:

$$r_p = \frac{m_p v_p}{qB} = \frac{(1.67 \times 10^{-27} \text{ kg})(4.8 \times 10^3 \text{ m/s})}{(1.6 \times 10^{-19} \text{ C})(0.002 \text{ T})} = 2.4 \times 10^{-2} \text{ m}$$

An electron will have a radius given by:

$$r_e = \frac{m_e v_e}{qB} = \frac{(9.11 \times 10^{-31} \text{ kg})(2.0 \times 10^5 \text{ m/s})}{(1.6 \times 10^{-19} \text{ C})(0.002 \text{ T})} = 5.7 \times 10^{-4} \text{ m}$$

The charges will each hit the wall one diameter's distance away so we can double the above two radii and see them strike the plate.

The proton strikes 4.8×10^{-2} m below the hole. The electron strikes 1.4×10^{-3} m above the hole. The charges will strike 0.049 m apart.

(d)   The time it takes to complete one circle would be given by $v = \dfrac{2\pi r}{T}$ if it traveled the full circle, but it only travels half way around so that we have

$$v = \frac{\pi r}{T}$$
$$T = \frac{\pi r}{v}$$

The proton will take

$$T_p = \frac{\pi(2.4 \times 10^{-2}\ \text{m})}{(4.8 \times 10^3\ \text{m/s})} = 1.6 \times 10^{-5}\ \text{s}$$

$$T_e = \frac{\pi(5.7 \times 10^{-4}\ \text{m})}{(2.0 \times 10^5\ \text{m/s})} = 9.0 \times 10^{-9}\ \text{s}$$

2.   (a)   A positively charged object will be repelled from the positive charge at position 0 and attracted to the negative charge as position $+d$, so it will experience a force to the right during the entirety of its motion. The object will remove from its release position directly toward the negative charge at position d with a velocity that increases during the entirety of the motion.

(b)   The field will be to the right in the two region $-\infty < x < -d$ and $0 < x < d$, so this is where the plot will be positive. The field magnitude will go toward infinity as the position approached the charges as $-d$, 0, and $+d$. The charge at 0 has a larger magnitude, so the positions where the charge has the smallest magnitude should be closer to the charges as $+d$ and $-d$.

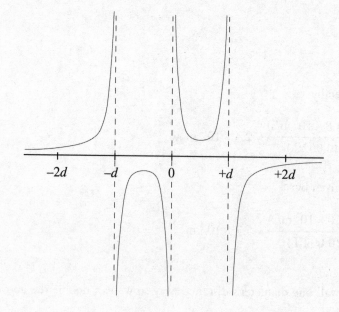

(c) In the region between 0 and d, the net charge will depend on the influence of each of the three charges on the x-axis.

$$E_{-d}(x) = \frac{kQ}{(d+x)^2} \text{ is directed to the left}$$

$$E_0(x) = \frac{k(2Q)}{(x)^2} \text{ is directed to the right}$$

$$E_{+d}(x) = \frac{kQ}{(d-x)^2} \text{ is directed to the right}$$

Thus, $E(x) = \dfrac{kQ}{(d-x)^2} + \dfrac{k(2Q)}{(x)^2} - \dfrac{kQ}{(d+x)^2}$ is directed to the right.

(d) The work required to bring a charge to a position is equal to the electric potential at that position multiplied by the amount of charge that is moved. Once the first two charges have been assembled, the electric potential at position $x = -d$ will be lower than when there was only a positive charge at $x = 0$ because the charge at $x = d$ is negative. Therefore, the work to bring in the third charge will be less than W.

3. (a) Any image that can be formed on a screen is a real inverted image.

(b) The thin lens equation states $\dfrac{1}{s_i} + \dfrac{1}{s_o} = \dfrac{1}{f}$. Picking the first two data points 15 and 61.5, we obtain

$$\frac{1}{s_i} + \frac{1}{s_o} = \frac{1}{f}$$
$$\frac{1}{61.5 \text{ cm}} + \frac{1}{15 \text{ cm}} = \frac{1}{f}$$
$$0.083 \text{ cm}^{-1} = \frac{1}{f} \rightarrow f = 12 \text{ cm}$$

if an image is placed 6 cm away from the lens we have

$$\frac{1}{s_i} + \frac{1}{s_o} = \frac{1}{f}$$
$$\frac{1}{s_i} = \frac{1}{f} - \frac{1}{s_o}$$
$$\frac{1}{s_i} = \frac{1}{12 \text{ cm}} - \frac{1}{6 \text{ cm}}$$
$$\frac{1}{s_i} = \frac{-1}{12 \text{ cm}} \rightarrow s_i = -12 \text{ cm}$$

This is a virtual upright image.

(c)   A ray-tracing diagram would look something like this.

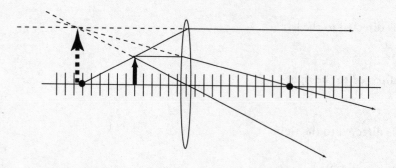

(d)   The diameter of the lens only affects the ability to gather light. It would decrease the brightness of the image but not change the location of the image.

4.   (a)

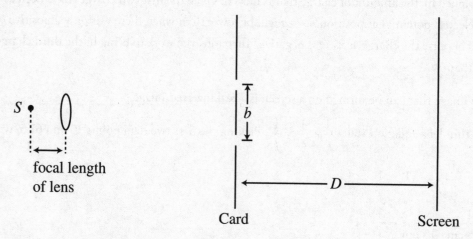

Note: Figure not drawn to scale.

(b)

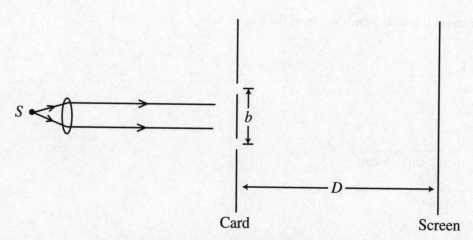

Note: Figure not drawn to scale.

(c)   In a double-slit experiment $x_m = \dfrac{m\lambda L}{d}$, where $x = L \sin \theta$. Using the corresponding quantities provided to us, we can change the equation to

$d \sin \theta = m\lambda$
$b \sin \theta = m\lambda$

Recall that $\sin \theta = y_3/D$,

$b\dfrac{y_3}{D} = 3\lambda$

$\lambda = \dfrac{y_3 b}{3D}$

(d)   The distance between the bright fringes would increase.

$n = \dfrac{c}{v} = \dfrac{\lambda_c f}{\lambda_m f} = \dfrac{\lambda_c}{\lambda_m}$

n must be less than 1, so

$\dfrac{\lambda_c}{\lambda_m} < 1 \Rightarrow \lambda_m > \lambda_c$ and $d \sin \theta = m\lambda$

thus, $\sin \theta$ would increase. Sin θ equals the ratio of the distance between the bright fringes to the distance between the card and the screen which is a constant. As a result, since $\sin \theta$ increases, the distance between the bright fringes would increase.

The
Princeton
Review®

Completely darken bubbles with a No. 2 pencil. If you make a mistake, be sure to erase mark completely. Erase all stray marks.

1.

YOUR NAME: _____
(Print)   Last   First   M.I.

SIGNATURE: _____   DATE: ___ / ___ / ___

HOME ADDRESS: _____
(Print)   Number and Street

_____
City   State   Zip Code

PHONE NO.: _____
(Print)

IMPORTANT: Please fill in these boxes exactly as shown on the back cover of your test book.

2. TEST FORM

6. DATE OF BIRTH

Month	Day		Year	
○ JAN				
○ FEB	⓪	⓪	⓪	⓪
○ MAR	①	①	①	①
○ APR	②	②	②	②
○ MAY	③	③	③	③
○ JUN		④	④	④
○ JUL		⑤	⑤	⑤
○ AUG		⑥	⑥	⑥
○ SEP		⑦	⑦	⑦
○ OCT		⑧	⑧	⑧
○ NOV		⑨	⑨	⑨
○ DEC				

3. TEST CODE   **4. REGISTRATION NUMBER**

⓪	Ⓐ	Ⓙ	⓪	⓪	⓪	⓪	⓪	⓪	⓪	⓪
①	Ⓑ	Ⓚ	①	①	①	①	①	①	①	①
②	Ⓒ	Ⓛ	②	②	②	②	②	②	②	②
③	Ⓓ	Ⓜ	③	③	③	③	③	③	③	③
④	Ⓔ	Ⓝ	④	④	④	④	④	④	④	④
⑤	Ⓕ	Ⓞ	⑤	⑤	⑤	⑤	⑤	⑤	⑤	⑤
⑥	Ⓖ	Ⓟ	⑥	⑥	⑥	⑥	⑥	⑥	⑥	⑥
⑦	Ⓗ	Ⓠ	⑦	⑦	⑦	⑦	⑦	⑦	⑦	⑦
⑧	Ⓘ	Ⓡ	⑧	⑧	⑧	⑧	⑧	⑧	⑧	⑧
⑨			⑨	⑨	⑨	⑨	⑨	⑨	⑨	⑨

7. SEX
○ MALE
○ FEMALE

The
Princeton
Review®

5. YOUR NAME

First 4 letters of last name				FIRST INIT	MID INIT
Ⓐ	Ⓐ	Ⓐ	Ⓐ	Ⓐ	Ⓐ
Ⓑ	Ⓑ	Ⓑ	Ⓑ	Ⓑ	Ⓑ
Ⓒ	Ⓒ	Ⓒ	Ⓒ	Ⓒ	Ⓒ
Ⓓ	Ⓓ	Ⓓ	Ⓓ	Ⓓ	Ⓓ
Ⓔ	Ⓔ	Ⓔ	Ⓔ	Ⓔ	Ⓔ
Ⓕ	Ⓕ	Ⓕ	Ⓕ	Ⓕ	Ⓕ
Ⓖ	Ⓖ	Ⓖ	Ⓖ	Ⓖ	Ⓖ
Ⓗ	Ⓗ	Ⓗ	Ⓗ	Ⓗ	Ⓗ
Ⓘ	Ⓘ	Ⓘ	Ⓘ	Ⓘ	Ⓘ
Ⓙ	Ⓙ	Ⓙ	Ⓙ	Ⓙ	Ⓙ
Ⓚ	Ⓚ	Ⓚ	Ⓚ	Ⓚ	Ⓚ
Ⓛ	Ⓛ	Ⓛ	Ⓛ	Ⓛ	Ⓛ
Ⓜ	Ⓜ	Ⓜ	Ⓜ	Ⓜ	Ⓜ
Ⓝ	Ⓝ	Ⓝ	Ⓝ	Ⓝ	Ⓝ
Ⓞ	Ⓞ	Ⓞ	Ⓞ	Ⓞ	Ⓞ
Ⓟ	Ⓟ	Ⓟ	Ⓟ	Ⓟ	Ⓟ
Ⓠ	Ⓠ	Ⓠ	Ⓠ	Ⓠ	Ⓠ
Ⓡ	Ⓡ	Ⓡ	Ⓡ	Ⓡ	Ⓡ
Ⓢ	Ⓢ	Ⓢ	Ⓢ	Ⓢ	Ⓢ
Ⓣ	Ⓣ	Ⓣ	Ⓣ	Ⓣ	Ⓣ
Ⓤ	Ⓤ	Ⓤ	Ⓤ	Ⓤ	Ⓤ
Ⓥ	Ⓥ	Ⓥ	Ⓥ	Ⓥ	Ⓥ
Ⓦ	Ⓦ	Ⓦ	Ⓦ	Ⓦ	Ⓦ
Ⓧ	Ⓧ	Ⓧ	Ⓧ	Ⓧ	Ⓧ
Ⓨ	Ⓨ	Ⓨ	Ⓨ	Ⓨ	Ⓨ
Ⓩ	Ⓩ	Ⓩ	Ⓩ	Ⓩ	Ⓩ

1. Ⓐ Ⓑ Ⓒ Ⓓ Ⓔ
2. Ⓐ Ⓑ Ⓒ Ⓓ Ⓔ
3. Ⓐ Ⓑ Ⓒ Ⓓ Ⓔ
4. Ⓐ Ⓑ Ⓒ Ⓓ Ⓔ
5. Ⓐ Ⓑ Ⓒ Ⓓ Ⓔ
6. Ⓐ Ⓑ Ⓒ Ⓓ Ⓔ
7. Ⓐ Ⓑ Ⓒ Ⓓ Ⓔ
8. Ⓐ Ⓑ Ⓒ Ⓓ Ⓔ
9. Ⓐ Ⓑ Ⓒ Ⓓ Ⓔ
10. Ⓐ Ⓑ Ⓒ Ⓓ Ⓔ
11. Ⓐ Ⓑ Ⓒ Ⓓ Ⓔ
12. Ⓐ Ⓑ Ⓒ Ⓓ Ⓔ
13. Ⓐ Ⓑ Ⓒ Ⓓ Ⓔ

14. Ⓐ Ⓑ Ⓒ Ⓓ Ⓔ
15. Ⓐ Ⓑ Ⓒ Ⓓ Ⓔ
16. Ⓐ Ⓑ Ⓒ Ⓓ Ⓔ
17. Ⓐ Ⓑ Ⓒ Ⓓ Ⓔ
18. Ⓐ Ⓑ Ⓒ Ⓓ Ⓔ
19. Ⓐ Ⓑ Ⓒ Ⓓ Ⓔ
20. Ⓐ Ⓑ Ⓒ Ⓓ Ⓔ
21. Ⓐ Ⓑ Ⓒ Ⓓ Ⓔ
22. Ⓐ Ⓑ Ⓒ Ⓓ Ⓔ
23. Ⓐ Ⓑ Ⓒ Ⓓ Ⓔ
24. Ⓐ Ⓑ Ⓒ Ⓓ Ⓔ
25. Ⓐ Ⓑ Ⓒ Ⓓ Ⓔ
26. Ⓐ Ⓑ Ⓒ Ⓓ Ⓔ

27. Ⓐ Ⓑ Ⓒ Ⓓ Ⓔ
28. Ⓐ Ⓑ Ⓒ Ⓓ Ⓔ
29. Ⓐ Ⓑ Ⓒ Ⓓ Ⓔ
30. Ⓐ Ⓑ Ⓒ Ⓓ Ⓔ
31. Ⓐ Ⓑ Ⓒ Ⓓ Ⓔ
32. Ⓐ Ⓑ Ⓒ Ⓓ Ⓔ
33. Ⓐ Ⓑ Ⓒ Ⓓ Ⓔ
34. Ⓐ Ⓑ Ⓒ Ⓓ Ⓔ
35. Ⓐ Ⓑ Ⓒ Ⓓ Ⓔ
36. Ⓐ Ⓑ Ⓒ Ⓓ Ⓔ
37. Ⓐ Ⓑ Ⓒ Ⓓ Ⓔ
38. Ⓐ Ⓑ Ⓒ Ⓓ Ⓔ
39. Ⓐ Ⓑ Ⓒ Ⓓ Ⓔ

40. Ⓐ Ⓑ Ⓒ Ⓓ Ⓔ
41. Ⓐ Ⓑ Ⓒ Ⓓ Ⓔ
42. Ⓐ Ⓑ Ⓒ Ⓓ Ⓔ
43. Ⓐ Ⓑ Ⓒ Ⓓ Ⓔ
44. Ⓐ Ⓑ Ⓒ Ⓓ Ⓔ
45. Ⓐ Ⓑ Ⓒ Ⓓ Ⓔ
46. Ⓐ Ⓑ Ⓒ Ⓓ Ⓔ
47. Ⓐ Ⓑ Ⓒ Ⓓ Ⓔ
48. Ⓐ Ⓑ Ⓒ Ⓓ Ⓔ
49. Ⓐ Ⓑ Ⓒ Ⓓ Ⓔ
50. Ⓐ Ⓑ Ⓒ Ⓓ Ⓔ

Completely darken bubbles with a No. 2 pencil. If you make a mistake, be sure to erase mark completely. Erase all stray marks.

1.

YOUR NAME: _____
(Print)

Last   First   M.I.

SIGNATURE: _____   DATE: ___/___/___

HOME ADDRESS: _____
(Print)

Number and Street

City   State   Zip Code

PHONE NO.: _____
(Print)

IMPORTANT: Please fill in these boxes exactly as shown on the back cover of your test book.

2. TEST FORM

6. DATE OF BIRTH

Month	Day		Year	
○ JAN				
○ FEB	⓪	⓪	⓪	⓪
○ MAR	①	①	①	①
○ APR	②	②	②	②
○ MAY	③	③	③	③
○ JUN		④	④	④
○ JUL		⑤	⑤	⑤
○ AUG		⑥	⑥	⑥
○ SEP		⑦	⑦	⑦
○ OCT		⑧	⑧	⑧
○ NOV		⑨	⑨	⑨
○ DEC				

3. TEST CODE

⓪	Ⓐ	Ⓙ	⓪	⓪
①	Ⓑ	Ⓚ	①	①
②	Ⓒ	Ⓛ	②	②
③	Ⓓ	Ⓜ	③	③
④	Ⓔ	Ⓝ	④	④
⑤	Ⓕ	Ⓞ	⑤	⑤
⑥	Ⓖ	Ⓟ	⑥	⑥
⑦	Ⓗ	Ⓠ	⑦	⑦
⑧	Ⓘ	Ⓡ	⑧	⑧
⑨			⑨	⑨

4. REGISTRATION NUMBER

⓪	⓪	⓪	⓪	⓪	⓪	⓪
①	①	①	①	①	①	①
②	②	②	②	②	②	②
③	③	③	③	③	③	③
④	④	④	④	④	④	④
⑤	⑤	⑤	⑤	⑤	⑤	⑤
⑥	⑥	⑥	⑥	⑥	⑥	⑥
⑦	⑦	⑦	⑦	⑦	⑦	⑦
⑧	⑧	⑧	⑧	⑧	⑧	⑧
⑨	⑨	⑨	⑨	⑨	⑨	⑨

7. SEX

○ MALE
○ FEMALE

The **Princeton** Review®

5. YOUR NAME

First 4 letters of last name				FIRST INIT	MID INIT
Ⓐ	Ⓐ	Ⓐ	Ⓐ	Ⓐ	Ⓐ
Ⓑ	Ⓑ	Ⓑ	Ⓑ	Ⓑ	Ⓑ
Ⓒ	Ⓒ	Ⓒ	Ⓒ	Ⓒ	Ⓒ
Ⓓ	Ⓓ	Ⓓ	Ⓓ	Ⓓ	Ⓓ
Ⓔ	Ⓔ	Ⓔ	Ⓔ	Ⓔ	Ⓔ
Ⓕ	Ⓕ	Ⓕ	Ⓕ	Ⓕ	Ⓕ
Ⓖ	Ⓖ	Ⓖ	Ⓖ	Ⓖ	Ⓖ
Ⓗ	Ⓗ	Ⓗ	Ⓗ	Ⓗ	Ⓗ
Ⓘ	Ⓘ	Ⓘ	Ⓘ	Ⓘ	Ⓘ
Ⓙ	Ⓙ	Ⓙ	Ⓙ	Ⓙ	Ⓙ
Ⓚ	Ⓚ	Ⓚ	Ⓚ	Ⓚ	Ⓚ
Ⓛ	Ⓛ	Ⓛ	Ⓛ	Ⓛ	Ⓛ
Ⓜ	Ⓜ	Ⓜ	Ⓜ	Ⓜ	Ⓜ
Ⓝ	Ⓝ	Ⓝ	Ⓝ	Ⓝ	Ⓝ
Ⓞ	Ⓞ	Ⓞ	Ⓞ	Ⓞ	Ⓞ
Ⓟ	Ⓟ	Ⓟ	Ⓟ	Ⓟ	Ⓟ
Ⓠ	Ⓠ	Ⓠ	Ⓠ	Ⓠ	Ⓠ
Ⓡ	Ⓡ	Ⓡ	Ⓡ	Ⓡ	Ⓡ
Ⓢ	Ⓢ	Ⓢ	Ⓢ	Ⓢ	Ⓢ
Ⓣ	Ⓣ	Ⓣ	Ⓣ	Ⓣ	Ⓣ
Ⓤ	Ⓤ	Ⓤ	Ⓤ	Ⓤ	Ⓤ
Ⓥ	Ⓥ	Ⓥ	Ⓥ	Ⓥ	Ⓥ
Ⓦ	Ⓦ	Ⓦ	Ⓦ	Ⓦ	Ⓦ
Ⓧ	Ⓧ	Ⓧ	Ⓧ	Ⓧ	Ⓧ
Ⓨ	Ⓨ	Ⓨ	Ⓨ	Ⓨ	Ⓨ
Ⓩ	Ⓩ	Ⓩ	Ⓩ	Ⓩ	Ⓩ

1. Ⓐ Ⓑ Ⓒ Ⓓ Ⓔ
2. Ⓐ Ⓑ Ⓒ Ⓓ Ⓔ
3. Ⓐ Ⓑ Ⓒ Ⓓ Ⓔ
4. Ⓐ Ⓑ Ⓒ Ⓓ Ⓔ
5. Ⓐ Ⓑ Ⓒ Ⓓ Ⓔ
6. Ⓐ Ⓑ Ⓒ Ⓓ Ⓔ
7. Ⓐ Ⓑ Ⓒ Ⓓ Ⓔ
8. Ⓐ Ⓑ Ⓒ Ⓓ Ⓔ
9. Ⓐ Ⓑ Ⓒ Ⓓ Ⓔ
10. Ⓐ Ⓑ Ⓒ Ⓓ Ⓔ
11. Ⓐ Ⓑ Ⓒ Ⓓ Ⓔ
12. Ⓐ Ⓑ Ⓒ Ⓓ Ⓔ
13. Ⓐ Ⓑ Ⓒ Ⓓ Ⓔ

14. Ⓐ Ⓑ Ⓒ Ⓓ Ⓔ
15. Ⓐ Ⓑ Ⓒ Ⓓ Ⓔ
16. Ⓐ Ⓑ Ⓒ Ⓓ Ⓔ
17. Ⓐ Ⓑ Ⓒ Ⓓ Ⓔ
18. Ⓐ Ⓑ Ⓒ Ⓓ Ⓔ
19. Ⓐ Ⓑ Ⓒ Ⓓ Ⓔ
20. Ⓐ Ⓑ Ⓒ Ⓓ Ⓔ
21. Ⓐ Ⓑ Ⓒ Ⓓ Ⓔ
22. Ⓐ Ⓑ Ⓒ Ⓓ Ⓔ
23. Ⓐ Ⓑ Ⓒ Ⓓ Ⓔ
24. Ⓐ Ⓑ Ⓒ Ⓓ Ⓔ
25. Ⓐ Ⓑ Ⓒ Ⓓ Ⓔ
26. Ⓐ Ⓑ Ⓒ Ⓓ Ⓔ

27. Ⓐ Ⓑ Ⓒ Ⓓ Ⓔ
28. Ⓐ Ⓑ Ⓒ Ⓓ Ⓔ
29. Ⓐ Ⓑ Ⓒ Ⓓ Ⓔ
30. Ⓐ Ⓑ Ⓒ Ⓓ Ⓔ
31. Ⓐ Ⓑ Ⓒ Ⓓ Ⓔ
32. Ⓐ Ⓑ Ⓒ Ⓓ Ⓔ
33. Ⓐ Ⓑ Ⓒ Ⓓ Ⓔ
34. Ⓐ Ⓑ Ⓒ Ⓓ Ⓔ
35. Ⓐ Ⓑ Ⓒ Ⓓ Ⓔ
36. Ⓐ Ⓑ Ⓒ Ⓓ Ⓔ
37. Ⓐ Ⓑ Ⓒ Ⓓ Ⓔ
38. Ⓐ Ⓑ Ⓒ Ⓓ Ⓔ
39. Ⓐ Ⓑ Ⓒ Ⓓ Ⓔ

40. Ⓐ Ⓑ Ⓒ Ⓓ Ⓔ
41. Ⓐ Ⓑ Ⓒ Ⓓ Ⓔ
42. Ⓐ Ⓑ Ⓒ Ⓓ Ⓔ
43. Ⓐ Ⓑ Ⓒ Ⓓ Ⓔ
44. Ⓐ Ⓑ Ⓒ Ⓓ Ⓔ
45. Ⓐ Ⓑ Ⓒ Ⓓ Ⓔ
46. Ⓐ Ⓑ Ⓒ Ⓓ Ⓔ
47. Ⓐ Ⓑ Ⓒ Ⓓ Ⓔ
48. Ⓐ Ⓑ Ⓒ Ⓓ Ⓔ
49. Ⓐ Ⓑ Ⓒ Ⓓ Ⓔ
50. Ⓐ Ⓑ Ⓒ Ⓓ Ⓔ

NOTES

NOTES

NOTES

NOTES

NOTES

NOTES

NOTES

International Offices Listing

China (Beijing)
1501 Building A,
Disanji Creative Zone,
No.66 West Section of North 4th Ring Road Beijing
Tel: +86-10-62684481/2/3
Email: tprkor01@chol.com
Website: www.tprbeijing.com

China (Shanghai)
1010 Kaixuan Road
Building B, 5/F
Changning District, Shanghai, China 200052
Sara Beattie, Owner: Email: tprenquiry.sha@sarabeattie.com
Tel: +86-21-5108-2798
Fax: +86-21-6386-1039
Website: www.princetonreviewshanghai.com

Hong Kong
5th Floor, Yardley Commercial Building
1–6 Connaught Road West, Sheung Wan, Hong Kong
(MTR Exit C)
Sara Beattie, Owner: Email: tprenquiry.sha@sarabeattie.com
Tel: +852-2507-9380
Fax: +852-2827-4630
Website: www.princetonreviewhk.com

India (Mumbai)
Score Plus Academy
Office No.15, Fifth Floor
Manek Mahal 90
Veer Nariman Road
Next to Hotel Ambassador
Churchgate, Mumbai 400020
Maharashtra, India
Ritu Kalwani: Email: director@score-plus.com
Tel: + 91 22 22846801 / 39 / 41
Website: www.scoreplusindia.com

India (New Delhi)
South Extension
K–16, Upper Ground Floor
South Extension Part–1,
New Delhi-110049
Aradhana Mahna: aradhana@manyagroup.com
Monisha Banerjee: monisha@manyagroup.com
Ruchi Tomar: ruchi.tomar@manyagroup.com
Rishi Josan: Rishi.josan@manyagroup.com
Vishal Goswamy: vishal.goswamy@manyagroup.com
Tel: +91-11-64501603/ 4, +91-11-65028379
Website: www.manyagroup.com

Lebanon
463 Bliss Street
AlFarra Building–2nd floor
Ras Beirut
Beirut, Lebanon
Hassan Coudsi: Email: hassan.coudsi@review.com
Tel: +961-1-367-688
Website: www.princetonreviewlebanon.com

Korea
945-25 Young Shin Building
25 Daechi-Dong, Kangnam-gu
Seoul, Korea 135-280
Yong-Hoon Lee: Email: TPRKor01@chollian.net
In-Woo Kim: Email: iwkim@tpr.co.kr
Tel: + 82-2-554-7762
Fax: +82-2-453-9466
Website: www.tpr.co.kr

Kuwait
ScorePlus Learning Center
Salmiyah Block 3, Street 2 Building 14
Post Box: 559, Zip 1306, Safat, Kuwait
Email: infokuwait@score-plus.com
Tel: +965-25-75-48-02 / 8
Fax: +965-25-75-46-02
Website: www.scorepluseducation.com

Malaysia
Sara Beattie MDC Sdn Bhd
Suites 18E & 18F
18th Floor
Gurney Tower, Persiaran Gurney
Penang, Malaysia
Email: tprkl.my@sarabeattie.com
Sara Beattie, Owner: Email: tprenquiry.sha@sarabeattie.com
Tel: +604-2104 333
Fax: +604-2104 330
Website: www.princetonreviewKL.com

Mexico
TPR México
Guanajuato No. 242 Piso 1 Interior 1
Col. Roma Norte
México D.F., C.P.06700
registro@princetonreviewmexico.com
Tel: +52-55-5255-4495
+52-55-5255-4440
+52-55-5255-4442
Website: www.princetonreviewmexico.com

Qatar
Score Plus
Villa No. 49, Al Waab Street
Opp Al Waab Petrol Station
Post Box: 39068, Doha, Qatar
Email: infoqatar@score-plus.com
Tel: +974 44 36 8580, +974 526 5032
Fax: +974 44 13 1995
Website: www.scorepluseducation.com

Taiwan
The Princeton Review Taiwan
2F, 169 Zhong Xiao East Road, Section 4
Taipei, Taiwan 10690
Lisa Bartle (Owner): lbartle@princetonreview.com.tw
Tel: +886-2-2751-1293
Fax: +886-2-2776-3201
Website: www.PrincetonReview.com.tw

Thailand
The Princeton Review Thailand
Sathorn Nakorn Tower, 28th floor
100 North Sathorn Road
Bangkok, Thailand 10500
Thavida Bijayendrayodhin (Chairman)
Email: thavida@princetonreviewthailand.com
Mitsara Bijayendrayodhin (Managing Director)
Email: mitsara@princetonreviewthailand.com
Tel: +662-636-6770
Fax: +662-636-6776
Website: www.princetonreviewthailand.com

Turkey
Yeni Sülün Sokak No. 28
Levent, Istanbul, 34330, Turkey
Nuri Ozgur: nuri@tprturkey.com
Rona Ozgur: rona@tprturkey.com
Iren Ozgur: iren@tprturkey.com
Tel: +90-212-324-4747
Fax: +90-212-324-3347
Website: www.tprturkey.com

UAE
Emirates Score Plus
Office No: 506, Fifth Floor
Sultan Business Center
Near Lamcy Plaza, 21 Oud Metha Road
Post Box: 44098, Dubai
United Arab Emirates
Hukumat Kalwani: skoreplus@gmail.com
Ritu Kalwani: director@score-plus.com
Email: info@score-plus.com
Tel: +971-4-334-0004
Fax: +971-4-334-0222
Website: www.scorepluseducation.com

Our International Partners

The Princeton Review also runs courses with a variety of partners in Africa, Asia, Europe, and South America.

Georgia
LEAF American-Georgian Education Center
www.leaf.ge

Mongolia
English Academy of Mongolia
www.nyescm.org

Nigeria
The Know Place
www.knowplace.com.ng

Panama
Academia Interamericana de Panama
http://aip.edu.pa/

Switzerland
Institut Le Rosey
http://www.rosey.ch/

All other inquiries, please email us at
internationalsupport@review.com